Architectural Design and Management in the Digital Age

Architectural Design and Management in the Digital Age

International Perspectives

Edited by
Bob Giddings
Northumbria University
UK

Silvio Melhado
University of Quebec
Canada

Jenni Barrett
University of Central Lancashire
UK

Working Commission 096,
Architectural Design and Management

WILEY Blackwell

This edition first published 2025

Registered Offices
John Wiley & Sons, Inc., 111 River Street, Hoboken, NJ 07030, USA
John Wiley & Sons Ltd, The Atrium, Southern Gate, Chichester, West Sussex, PO19 8SQ, UK

For details of our global editorial offices, customer services, and more information about Wiley products visit us at www.wiley.com.

Wiley also publishes its books in a variety of electronic formats and by print-on-demand. Some content that appears in standard print versions of this book may not be available in other formats.

Library of Congress Cataloging-in-Publication Data Applied for:

Hardback ISBN: 9781119902294

Cover Design: Wiley
Cover Image: City Hall, London, Courtesy of Foster + Partners

Set in 9.5/12.5pt STIXTwoText by Straive, Chennai, India

Printed and bound by CPI Group (UK) Ltd, Croydon, CR0 4YY

C9781119902294_230824

Contents

About the Editors

Bob Giddings: Emeritus Professor of Architecture and Urban Design in the Department of Architecture and Built Environment, Northumbria University. He holds a DPhil from the Institute of Advanced Architectural Studies, University of York, the UK. Bob has supervised over 20 PhDs to successful conclusion. He has been involved in authoring eight books and numerous academic journal and international conference papers. He initiated the Architecture Programme at Northumbria University and was the Lead Examiner at the Architects Registration Board for many years. He is a partner in the renowned Future of the City Centre Research Network. Bob is currently the Joint Co-ordinator of CIBW096.

Silvio Melhado: Professor of Construction Management at the University of Quebec, École de technologie supérieure and Senior Professor at the University of São Paulo, Escola Politécnica. He holds a PhD in design management from Brazil, and a postdoctoral qualification in social sciences from France. Silvio has supervised over 200 graduate students and has published more than 350 academic journal and conference papers. He has also authored 20 book chapters and eight books. He is a member of the Ordre des Ingénieurs du Québec and a former Co-ordinator of the Brazilian National Group of Design Quality. Silvio is currently the Joint Co-ordinator of CIBW096.

Jenni Barrett: Principal Lecturer in the School of Business at the University of Central Lancashire. She holds a PhD from the School of Psychology at the University and a master's degree in Interdisciplinary Design for the Built Environment from the University of Cambridge. Jenni has supervised numerous postgraduate theses relating to the impacts of technology on organisational culture and creativity, and has published in academic journals and at international conferences. She has taught design and management across architectural programmes and was Programme Leader for the Master's in architecture course. She is also a member of Global Executive Team for the international network, Women In BIM.

About the Contributors

Monia Del Pinto: Vice-Chancellor Independent Research Fellow in the School of Architecture, Building, and Civil Engineering at Loughborough University, UK. Monia holds a PhD in urban planning and disaster risk reduction from Loughborough University and an MSc in architectural engineering from Universita' degli Studi di L'Aquila, Italy. She works at the intersections of architecture, urban planning, heritage studies and disaster management. Her multidisciplinary research aims at innovating tools for effective disaster risk reduction at an urban scale. In her current study, Monia is developing a global methodology to assess vulnerability to climate-induced hazards associated with the spatial networks in heritage cities.

Stephen Emmitt: Professor of Architectural Practice at the University of Bath, Department of Architecture and Civil Engineering. Stephen joined CIBW096 in 1994, while still an architect in practice. He served as press officer from 1996 to 2004 and held the role of Joint Co-ordinator from 2005 to 2011, responsible for organising conferences, editing conference proceedings and providing direction to the working commission. He received a CIB 2009 Programme Committee Commendation for outstanding contribution to CIBW096's mission. He has written extensively on architectural management and design management. His most recent book is *Design Management for Architects*, Third Edition, 2024.

Fang Zhou: Lecturer at the School of Architecture and Urban Planning, Huazhong University of Science and Technology, Wuhan, China. She was awarded BA in environmental design and BEng in architecture by the University, MSc in architectural studies by the University of Florida and PhD in building and real estate by the Hong Kong Polytechnic University. Zhou's research interests include design management and design methodology. Her work has been funded by the Natural Science Foundation of Hubei Province in China, the Fundamental Research Funds for Central Universities in China and a PhD Fellowship in Hong Kong. Zhou is a member of CIB096.

Mary Hardie: Associate Professor in Building Design Management and Architecture at Western Sydney University. Mary has been an academic contributing to teaching and research in construction, design and architecture since the 1990s. Her work focuses on enabling diverse cultural perspectives to find welcoming pathways into satisfying careers in the built environment. She has written extensively on educational perspectives in construction as well as the fostering of innovation and cultural change in industry. CIB W096 Architectural Design and Management has provided her with an excellent forum to discuss the future of the industry with international colleagues and researchers.

Ivanka Iordanova: Associate Professor in the Department of Construction Engineering at ÉTS, Montreal, and a member of GRIDD. After a career as the Director of Innovation at a large general contractor, her teaching and research now focuses on digital innovation and new construction strategies aiming at the realisation of high-quality buildings and infrastructures, through BIM/VDC, Lean Construction and Robotic Fabrication. With a goal of increasing construction productivity, quality and safety, she studies the potential of industrialisation of construction and digital technologies for making the built environment more sustainable, harmonious and empathetic. She is a founding member of Lean Design and Construction Canada.

Aya Jazaierly: Lecturer in Architecture at Loughborough University, the United Kingdom. She is involved in both BArch and MArch programmes, teaching design studio and professional practice. Aya was awarded a PhD by the University of Tokyo. She worked in Arata Isozaki's office, and as part of Kengo Kuma's Research Laboratory at the University of Tokyo between 2012 and 2019. She has published on the philosophy of Japanese architects. Her research interests are focused on the design process. This pertains to all the tangible and intangible creative, sociological, political and psychological practices that define the intellectual, formal and aesthetical manifestations of architectural and urban output.

Gabriel Jobidon: Associate Professor in the Department of Construction Engineering at ÉTS and member of GRIDD. He holds a PhD in law from Université Laval, Quebec City, where he studied in the Chaire industrielle de recherche sur la construction écoresponsable en bois. His multidisciplinary research interests include public procurement, collaborative contracts, and adaptation to climate change. He has written several book chapters, academic journal and conference papers on topics including integrated project delivery and collaborative contracting strategies. Gabriel has worked on major infrastructure projects in the Province of Quebec and continues to act as a consultant to public organisations.

Linda Kestle: Associate Professor in Construction at Unitec Institute of Technology, Auckland, New Zealand. She holds a PhD in remote site design management from the University of Canterbury, New Zealand, which involved design management modelling for multinational collaborative projects. She is a Fellow of the NZIOB and a Chartered Building Professional. Her long career includes research publications in international conferences and journals, architectural and engineering design consultancy and on-site construction project management. Linda is a member of the CIBW096, and her current research interests include academia–industry research partnerships, digitalisation of construction industry practice and education, the circular economy and sustainable design.

Ania Khodabakhshian: Postdoctoral Researcher at the B John Garrick Institute of Risk Sciences, UCLA. Ania holds a PhD with honours in architecture, built environment and construction engineering from Politecnico di Milano. Her research focuses on the application of artificial intelligence in construction projects, risk management in engineering systems and data-driven building stock energy retrofits. Her work has achieved significant recognition through international accolades such as the CIB Sebestyen Future Leaders' award and Vistas funding, and publications in leading scientific journals and conferences. She is a member of CIB096, and she collaborates with renowned universities such as MIT, Chalmers Institute of Technology and the University of Salford.

Urša Komac: Senior Lecturer in Architecture at Western Sydney University. She is a registered architect from Europe and holds a BA in architecture and MSc in landscape architecture from the University of Ljubljana and a PhD from the Barcelona School of Architecture. She has worked as an architect and as an architectural educator for more than 20 years in Slovenia, Spain, the United States and Australia. Urša is a member of CIB096, the Chamber of Architecture and Spatial Planning of Slovenia, and is an AACA standing panel member. Her research interests are architecture and the city, public space, heritage and architectural education.

William McKee: Associate Lecturer at the University of Sydney. William is a registered architect in New South Wales, who has studied in Scotland, France and at the Architectural Association in London, where he gained his master's degree as student of the year, and was nominated for the RIBA President's Medal for his dissertation. He has practised in London, and currently in Sydney as the Director at Common Office, which was founded on the design of interiors, buildings, urbanism and territory. His current work includes furniture, houses, community buildings and research initiatives. All projects are treated as equal within a rigorous design process.

Marina Machado: International BIM Project Manager. With a background in architecture from her own studio in Brazil, Marina transitioned her expertise into driving digital transformation within the construction realm. Her journey led to Europe, where she spearheaded BIM initiatives, aligning organisational strategies with ISO 19650 standards. Marina's leadership extended to managing BIM aspects on multimillion-euro construction ventures across diverse sectors such as education, pharmaceuticals and data centres worldwide. She gained a master's degree in BIM and integrated design from the University of Salford, the United Kingdom. Marina has spoken at numerous international conferences and published articles elucidating BIM implementation for design to manufacture workflows.

Leonardo Manzione: Director of Coordenar, a specialist BIM consultancy and design co-ordination company. Leonardo is also a lecturer on the postgraduate course in Construction Project Management on the Poli-Integra program at the University of São Paulo, Escola Politécnica, Brazil. He holds M.Sc. and PhD in Civil Engineering from the same University. His research and teaching focus mainly on design planning and co-ordination, design management and Building Information Modelling implementation. He has published several articles in scientific journals and at conferences, as well as technical guides and books. He is a member of the committee responsible for the Brazilian version of ISO 19650 standards.

David Morton: Associate Professor at Northumbria University School of Architecture, where he also earned his PhD. His practice work includes award-winning schemes in the United Kingdom and internationally. He has published pedagogic research on design processes, use of CAD and BIM in architectural design and the effects of digital methods on studio projects. David has developed learning outside the teaching space by the use of podcasts and QR codes to reinforce the subject area and educational focus. He received a best lecturer award at the University, which led to his current pan-departmental role, focusing on mixed-methods approaches to learning typologies.

Robert Osei-Kyei: Senior Lecturer in Construction Management and Building Design at School of Engineering, Design and Built Environment at Western Sydney University. He has published research output in one textbook on international PPPs, eight book chapters, more than 120 high-impact journal articles and six industry reports, with 5000 citations. He has been ranked among the world's top 2% most-cited scientists by Elsevier BV and Stanford University. Robert is a Fellow of the UK Higher Education Academy, and a member of the CIB PPP Task Group, Australian Institute of Building, Australian Institute of Project Management and International Centre for Complex Project Management.

Erik A. Poirier: Associate Professor in the Department of Construction Engineering at École de technologie supérieure (ÉTS) in Montreal. He heads the Research Group for Integration and Sustainable Development in the Built Environment (GRIDD). Erik specialises in digital transformation initiatives in the construction industry. He has published several papers on the process of BIM adoption and implementation and has assisted several government bodies across Canada in the digital transformation process. He also specialises in the integration and optimisation of information flows across value chains in design processes, as well as in information lifecycle management and the standardisation of digital practices in the built environment.

Regan Potangaroa: Professor at Massey University School of the Built Environment, Auckland, New Zealand. He recently earned the Te Rangaunua Hiranga Māori Award from the NZ Royal Society reflecting his lifelong commitment to humanitarian deployments, emergency responses and marae restorations, with over 200 deployments in 22 countries since 1996. A steadfast advocate for sustainable built environments, Regan seamlessly integrates his research into community projects. His exemplary work extends to international collaborations, such as with the United Nations and the International Federation of Red Cross and Red Crescent Societies, showcasing his impact on global resilience and well-being. Regan is a member of CIBW096.

Taija Puolitaival: Head of Civil Engineering programmes at the Faculty of Built Environment, Tampere University, Finland. She has a long career in construction management education working in New Zealand and in Finland. Her research focuses on digital technologies, related competences and their development in architecture, engineering and construction. Alongside her role as a senior lecturer in New Zealand, Taija worked at the national level as a chair of the BIM Education NZ group and as a member of the BIM

Acceleration Committee to assist the industry in adopting and implementing Building Information Modelling and integrating it as part of AEC education.

Noha Saleeb: Associate Professor in Creative Digital Technologies and Construction at Middlesex University. She is a consultant in sustainability, digital transformation, digital twins and AI. She is a director of BIM Africa, a member of the steering committee for the London Digital Twin Research Centre, BIM4Heritage, and the CIOB Innovation Group. Noha is the Editor-in-Chief of the *Smart Infrastructure and Construction* journal. She has authored more than 100 papers in journals, at conferences, in books and in industry reports. She has received the Hong Kong Institute of BIM award and the best European Women in Construction Award. Noha is a member of CIB096.

James Sheppard: Board Director at Equals Consulting, a project and cost management consultancy. James qualified and practised as an architect prior to becoming a project management surveyor and member of the Royal Institution of Chartered Surveyors. He has project managed high-profile, complex projects such as Transforming Tate Britain in 2013 and a major refurbishment of the Imperial War Museum in 2014. He has over 20 years' experience across education, residential and commercial sectors. In 2020, James matriculated onto the MSt Interdisciplinary Design Built Environment at the University of Cambridge, completing research into the impact of digital design processes on creative collaboration.

Katya Veleva: EMCC Senior Practitioner accredited leadership coach. Katya is a seasoned Equity, Diversity and Inclusion leader spanning architecture, construction and beyond. They passionately foster inclusive environments, driving collaboration and innovation. Their achievements include designing global mentoring programs and supporting industry giants such as Grimshaw Architects and Google. Katya's expertise in healthcare design and building information modelling has propelled digital adoption and diversity initiatives. Recognised globally for their impactful work, Katya has addressed audiences at AU Las Vegas and was honoured as one of the top 4 Construction Role Models at UK Construction Week. Their commitment to diversity and inclusion shapes inclusive workplaces worldwide.

Maria Vogiatzaki: Professor and the Head of School of Architecture and Planning at Anglia Ruskin University, the United Kingdom. She holds a PhD from the University of Bath on the design process of nonstandard structures. She is an educator, researcher, registered architect, principal fellow of the Higher Education Academy and editor-in-chief of the open-access peer-reviewed journal *ArchiDOCT*, which publishes ongoing doctoral research in architecture and is indexed in Scopus. Throughout her teaching career, Maria has emphasised the critical importance of creativity as a tool that enables thinking against fixed binaries, confidently envisaging the future and teaching tomorrow today. She is a member of CIBW096.

Preface

Technology is a juggernaut that does not have a reverse gear. Nevertheless, it is incumbent on academics and practitioners that it should not be accepted without critical appraisal. The intention of this book is to present the advantages of digital technology in appropriate circumstances but also to appreciate that alternative methods may continue to provide considerable benefits in other situations. The book reflects on the rapid and prolific development of technologies used in architectural design and management. It begins with an overview of digital and cloud-based tools, with Building Information Modelling as a key development. Applications such as big data, artificial intelligence and robotics are also evolving significantly. There is a wide range of current and future possibilities for their application within architectural projects during the design process, but also as ways to facilitate architectural management. However, the application of digital technology should be determined by the human context, different types of buildings and spaces, their materials and construction, and the various concepts of place across the globe. Thus, there is a danger of digital stereotyping. It is perhaps the lack of flexibility and over-complexity in operational methods and regulatory compliance that has contributed to observations about the technologies not being adopted as much as their advocates would have hoped. Output from first-world countries can be applied to other countries by those who do not share their traditions, cultures and work environments, especially in the case of disadvantaged and disenfranchised people. Yet, it is not to say that those in emerging economies cannot take part in the digital revolution.

Inclusion, equity and diversity enrich environments, especially when they involve small contributions by many people over long periods of time. The waxing and waning of standardisation and the controversial impact on design context are explored. For some, there is a growing interest in information management. Nevertheless, concerns continue to revolve around the possibility of diminishing architectural quality. It may be that digital tools are more relevant to building production. Digitalisation of design and construction has the potential to transform not only the way projects are managed but also the role of the architect within the design process. Without due consideration of the human and professional contexts, it can push the communication of design ideas towards complex data management rather than engagement with space, form and structure. This re-conceptualisation may necessitate a new understanding of the role of designers and their responsibilities in conjunction with emerging disciplines, such as the BIM manager, in the new technologically driven project environment. Yet, there needs to be an evaluation of these roles to determine if creativity is being replaced by data manipulation or whether they are being introduced to cooperate and coexist with established ones to form inclusive and socially responsible design teams. Finally, new digital technologies are creating a diverging vocabulary, which involves new concepts and definitions that require interpretation, especially with regard to experiences of the design process in different countries.

Bob Giddings
Silvio Melhado
Jenni Barrett

Part I

Context

1

Introduction

Stephen Emmitt

University of Bath, UK

Context

CIBW96 Architectural Management was established in 1993, taking its name from an international conference held at the University of Nottingham a year earlier. The working group was created as a forum for practitioners and researchers to meet, share and develop knowledge on aspects of management that applied to architectural businesses and architectural projects. The aim was to facilitate the uptake of management by architects as a means of staying competitive and as a vehicle to better influence the realisation of high-quality architecture. At the time, there was a feeling that architects were not adequately grounded in management during their education and practice – a sentiment that may still apply today. This lack of managerial acumen has and continues to pervade many architectural businesses, often leading to problems with service provision and lack of influence over the control of design output. The group provided a unique focus for architects and academics, who were keen to explore the value of management to the architectural profession, with the hosting of regular international conferences that encouraged debate and the sharing of knowledge. W96 published peer-reviewed articles in its journal, the *International Journal of Architectural Management*, along with edited conference proceedings of its meetings and joint meetings with other CIB working groups and task groups. The membership of the working group has always been dynamic and has evolved since its inception over 30 years ago, with members and contributors drawn from a very broad range of geographical locations. Similarly, the joint coordinators have changed over time, which has resulted in changes of emphasis and a wide variety of views on management.

Writing in 1995, Freling posed the question as to whether architectural management (AM) is a discipline or a specialist interest (Freling 1995). The question was deliberately left unanswered. In many respects, as the review of the literature presented later in this chapter indicates, architectural management has remained a niche interest, even though management is inherent to the daily work of an architect. The golden age of architectural management ran from the early 1990s into the mid-to-late 2000s, and as the attention shifted

Architectural Design and Management in the Digital Age: International Perspectives, First Edition.
Edited by Bob Giddings, Silvio Melhado, and Jenni Barrett.

towards digitalisation, building information modelling (BIM) and information communication technologies, interest waned. Bi-annual conferences were replaced by annual conferences and the *International Journal of Architectural Management* ceased operation. This resulted in academics publishing in other peer-reviewed journals, including *Architectural Engineering & Design Management*, which was initiated and launched by members of CIBW96 and TG49 Architectural Engineering. Fashions change, and like the golden age of management in the 1960s, the output peaked and then declined. The term architectural management reflects other knowledge domains, such as construction management, project management and facilities management, all of which have become established. Architectural management has never become established as the term to use, with many authors preferring the more established terms such as office management, practice management, job running and design management.

It was inevitable that the emerging working commission, CIBW096, would not only widen its remit at some point but also change its name to associate more with what was happening in architectural education, practice and research around the globe. The change of name to Architectural Design and Management CIBW096 in 2016 signalled a significant development in the evolution of the commission. Architectural design and management is an expansion of the original architectural management brief. This is made clear in the *CIB Architectural Design and Management Research Roadmap* (Giddings and Melhado 2019), which aligns the commission with international architectural research and highlights a number of themes for the future. The initial categories in the Research Roadmap are design, technology and environment, cultural contexts, communities and digital technologies, and management, practice and law. This differs from the early work of CIBW96 that saw architectural management as central to being an architect in practice (Nicholson 1995). In the early years of the group's existence taking a holistic view to management was seen as a strength, but as time progressed, it started to become evident that a more structured and focused approach was required if architectural research was to develop, a point made explicit by Alharbi (2013).

The Research Roadmap separates the management aspects, which is what the early pioneers of architectural management would have seen as antithetical to their views. However, the change of name and focus makes the group more attractive to a wider audience and reflects the way in which management is seen by the international architectural research community. It also allows the working commission to build working relationships with international schools of architecture. As a result of the change to Architectural Design and Management CIBW096, it has increased its activities with these international schools of architecture. For example, it has established a relationship with the Architectural Research Centers Consortium, which is a group of American schools of architecture. It has also initiated a city centre sustainable design housing competition in 2023, with eight schools of architecture in South Africa, for which there were 68 entries. These are activities that are facilitated by integrating architectural design with management. The Roadmap also sets out a vision for the commission as a research entity that could lead and partner research funding applications. This is a worthy goal and a positive departure from the original remit of the group, which was mainly concerned with research that advanced the field for the benefit of practitioners. Architectural management is an area in which securing national and international research funding is notoriously difficult, and the broader remit of W096 better positions the commission to respond to funding opportunities compared with its earlier incarnations.

In this chapter, the term architectural management is used primarily to relate to the history of the working group and the knowledge domain in general. The term architectural design and management is also used, reflecting the current title of CIBW096 and its wider reach. Readers are encouraged to interpret the terminology to suit their own situations. The aim of this chapter is to provide background to the heritage of architectural management by way of a reflective commentary. This sets the scene to briefly revisit the four components of the architectural management framework as a way of providing context to the contents of this book. The chapter concludes with a call for integration rather than more management and posits that digital tools are one way of helping architects achieve better synergy between architectural design and management.

Background to Architectural Management

Management Applied to Architectural Practice by Brunton et al.'s (1964) is the first book to use and define the term architectural management. This publication was a response to the RIBA's critical report on the state of architectural practice in the 1960s. *The Architect and His Office* (RIBA 1962) had identified the need for better management skills, especially in relation to the management of the architectural business. Brunton et al. provided a sound argument in their book for applying management and stated that

> *. . . architectural management falls into two distinct parts, office or practice management and project management. The former provides an overall framework within which many individual projects will be commenced, managed and completed. In principle both parts have the same objectives but the techniques vary and mesh only at certain points (Brunton et al. 1964: p. 9)*

Although computing was in its infancy, Brunton et al. recognised the importance of computers and mechanisation of processes, such as for project programming, costing, and user studies of buildings. They were, however, keen to make the point that computers could not (at the time) make valuable judgements, unlike architects.

There was considerable interest in management in the 1960s by the architectural profession, although it waned during the 1970s as the architectural profession retreated from business management. Changing market conditions, especially the uptake of design management by the main contractors and the emergence of the independent project manager, led to greater attention to management in the late 1980s and early 1990s (see Emmitt 1999 for a detailed review). Renewed interest was reflected in Sharp's book, *The Business of Architecture* (1986), which extolled the value of business to the architectural profession. It was only a matter of time until someone dedicated an entire conference to the subject – the genius of what was to become CIBW96 (now CIBW096). Selected proceedings of the first conference on architectural management were published in an edited book by the conference organiser and initial coordinator of CIBW96, Paul Nicholson. *Architectural Management* (Nicholson 1992) was the first edited book to carry the name architectural management in its title and was endorsed by the CIB. It was not, however, the output of CIBW96, as the working group was established afterwards

(Nicholson and Prins 1993). In the absence of relevant publications, Nicholson (1992) provided an important point of reference for CIBW96 members and stimulated much of the early research articles that were subsequently published in the *International Journal of Architectural Management*. Although the book has many strengths, it could be criticised for not defining architectural management and for not drawing on the earlier work of Brunton et al. (1964). Nevertheless, the book contents cover a broad range of subject areas that included architectural management and practice, design management, project management, facilities management, value engineering and quality, computing in architecture, education and human resources. The digital theme was addressed in chapters on data transfer, IT use, information management, computer-aided drafting (CAD) utilisation and coding (Nicholson 1992).

Architectural Management: A Competitive Approach (Emmitt 1999) was the first monograph to carry the term architectural management in its title. The book was grounded in the practical experience of the author and the research literature. It was an attempt to promote the benefits to practitioners at a time when architects were facing significant competition for services from project managers and contractors due to the rapid uptake of design and build procurement (Allinson 1993; Gray et al. 1994). For the first time, there was an attempt to define architectural management in a book. This was done by drawing on the seminal work of Brunton et al. (1964), research on the workings of architectural practice (Blau 1984; Cuff 1991; Symes et al. 1995), the knowledge produced by CIBW96 and integrating the ideas from parallel research, such as the management of professional service firms (Maister 1993). The book was unique in that it addressed the management of the business and the management of projects for the first time in a monograph. Prior to this publication, the two components of architectural management had been tackled separately, either as practice management or project management. The digital theme was addressed in the book mainly through attention to the management of information within the office and within project networks.

The first book to be published by CIBW096 Architectural Management was an edited book of chapters written by members of the working group. The book *Architectural Management: International Research & Practice* (Emmitt et al. 2009) contains a mixture of chapters and case studies that explore aspects of valuing design, communication of design values, design management, inclusive design, integral design and innovation, architectural practice and education. Some of the themes have parallels with the edited book by Nicholson (1992), although the content of the book demonstrates a greater degree of maturity in thinking about what architectural management is and how it was being applied at the time. This was an important publication for W096, reflecting the international expertise within the working group, and the wide interpretation of the subject area signposted inter-related knowledge areas for further exploration and development. It is interesting to note in the introduction of the book that there is an argument for maintaining the term architectural management in preference to architectural design management, with design management seen as a subset of architectural management based on the earlier work into taxonomy (Emmitt et al. 2009). The digital theme was included in chapters on communication, project webs and creative software technologies associated with freeform design (3-D CAD and BIM). It should also be noted that climate change and the role of building users are featured in the book, highlighting the value of management to architects in a changing environment. Implicit in the edited work is a tension between the need to take a holistic view as argued by Nicholson (1995) and the separation into sub-themes, as has been done more recently in the CIBW096 Research Roadmap (Giddings and Melhado 2019).

The digital theme was further addressed in two CIBW096 conferences, one held in Taiwan (Tzeng et al. 2009) and one in Austria in 2011 (den Otter et al. 2011). The Taiwan conference was concerned with advancing the architectural management field and contributions to the conference included the integration of digital technologies with architectural businesses and their projects. This sets the theme for the Austria conference, which took the digital theme further, dedicating the entire conference to the development of digital technologies and their application to architectural businesses (den Otter et al. 2011). In many respects that conference was instrumental in re-orientating the working group to encompass digital architectural practice, a theme that has been continued with this book.

The definition of architectural management was revisited and redefined in a doctoral thesis by Alharbi (2013). He was the first to engage in extensive empirical research, which was informed by interactions with architects and the CIBW096 community via its conferences. The thesis drew on the knowledge base produced by CIBW096 and more widely on the literature dedicated to architectural design and management. This helped to inform the research and shape the understanding of architectural management in a competitive and collaborative marketplace for professional services. The thesis and associated research papers set out a new definition that was reached through consensus and a comprehensive framework that included four components of architectural management grounded in empirical research. The definition from the thesis was

> *Architectural Management (AM) is the strategic management of the architectural firm that assures the effective integration between managing the business aspects of the office with its individual projects in order to design and deliver the best value to all stakeholders. (Alharbi 2013: 148)*

The two components of architectural management were expanded from the earlier understanding used by Brunton et al. (1964) and Emmitt (1999) to include four components. These were managing the business, managing projects, managing stakeholders and managing learning. The rationale behind the themes and the expansive framework can be found in the thesis (Alharbi 2013). Although the research was conducted at a time of great interest in BIM, he chose to concentrate on management principles, arguing that the digital tools would evolve, whereas the principles of management as applied to architectural practice would not. Brunton et al. (1964) made a similar point about management principles having longevity and recognising that the computing aspects would evolve. The thesis formed the research background to a book, the *Handbook for the Architectural Manager* (Emmitt and Alharbi 2018). The book was written to provide a unique practical guide for architects and designed to be a source of inspiration, rather than how to do it, given that readers' contexts would differ. The digital theme is included in discussion of information and communication technologies (ICTs) and the use of information technology (IT) systems.

Since the publication of the *Handbook for the Architectural Manager* (Emmitt and Alharbi 2018), there have been significant technological advances in digital technologies and software, which are covered extensively in this book. There have also been significant changes in society, the demands of clients and the expectations of architects and fellow professionals as employees. Curiously, the literature on architectural management appears to have stalled at a time when it could be more relevant. There are a small number of more recent research articles related to management in architectural design, for example by Bondare et al. (2017), which reinforces the call for architects to pay more attention to management and Nikseresht et al. (2022), which supports the findings of earlier research by Alharbi (2013). Apart from the CIBW096 Research Roadmap (Giddings and Melhado 2019) that signposts new lines of enquiry, there is no evidence of new insights into architectural management theory, nor is there any evidence of recent critical debate about architectural management in the peer-reviewed journals. A similar criticism could be levelled at the design management literature and the literature on practice management. Perhaps there is a feeling that everything has already been said and done, and researchers have moved onto other areas of interest, such as digital tools, building performance and addressing the climate emergency. The reason may also be related to research funding, which, as noted above, is challenging to acquire for this subject area, and without funding it is increasingly difficult to find the time to dedicate to unfunded research. Another reason for the apparent decline in interest could relate to what is happening in architectural practice. Architects have been employing architectural managers and design managers, although they are not always using these terms for the job descriptions. More recently, they changed direction to BIM and data managers, who do not necessarily have any education in architectural design. While this tends to be associated with the larger architectural businesses, it is slowly diffusing to the medium-sized organisations, which are adopting digital tools and new ways of working. For example, the implementation of BIM and the associated BIM protocols has brought about more structured and better data management, which helps to improve workflow within the architectural business. Changes are happening in architecture, and the underlying principles of design quality and management need to be upheld in the digital age.

Fifteen years have passed since the first edited book was published by CIBW096, and much has changed in relation to the digital tools available to architects in that time. Much has also changed in relation to the urgent need to tackle climate change through better-performing buildings, which may be facilitated using digital technologies and the application of design and management processes architecture. Computing and digital technologies have always been present in the architectural design and management literature, although it has often dealt with implicitly as part of a wider agenda. In revisiting the earlier publications on architectural management, it is easy to see the attention shifting from computers in general, to CAD and information management, to BIM and building information management, through to generative design and automation. Most of the earlier research work was concerned with efficiency and accuracy of information production, dissemination, storage and access. It is only more recently that the opportunities afforded by generative artificial intelligence (AI) and machine learning have started to be discussed and implemented. The change to architectural practice may be profound.

Exploring the Components of Architectural Design and Management in the Digital Age

As a means of providing context to the contents of this book, it is insightful to briefly explain the four themes of architectural management, with links made to the chapters that follow through the section headings of the book. The intention is to explore the issues in relation to current situation for architects around the world. Readers are encouraged to interpret the issues in relation to the changing nature of working as an architect in different geographical contexts. These aspects, along with national and international laws, local cultures and traditions, and the maturity, availability and use of digital technologies will colour how architectural design and management are interpreted and applied. Architectural businesses are people-centric professional service firms. Regardless of size, their primary function is to create and deliver design value to their clients, and more widely to society. They have four distinct characteristics – creative, professional, dependent on the construction sector for business and service providers. Architectural businesses require a proactive, adaptive and engaged approach to design and management that relies on clear leadership, use of appropriate digital technologies and engaged people (Emmitt 2024). As knowledge-based organisations, architectural practice has always relied on human talent to create architectural design solutions. This can be facilitated by using software to help with the creative act of designing and information production and management of the design process. This helps with efficiency, accuracy and creativity. The four components of architectural management are as follows.

Managing the Business

From a business perspective, digital tools and technologies are increasingly being seen as central to a well-managed architectural business, facilitating hybrid working, collaboration and effective mixed-mode project teams. This is contributing to a transformation in architectural management culture. The automation of mundane administrative and drawing tasks has also freed up design talent to dedicate a greater percentage of their time to creative and entrepreneurial endeavours. The rapid uptake of generative design and AI has also started to raise questions about the type of employee architectural businesses need, with some businesses restructuring their business and employing architects with a wider skills base than would traditionally be associated with architectural design. Opportunities for further research relate to how architectural businesses harness the power of digital technologies to further the advancement of architecture and the health and well-being of their staff. It is, however, the sociological aspects of architectural businesses that remain largely unexplored, including leadership and governance in a digital work environment. One is left wondering what a sociological exploration of modern digital architectural practice conducted in the spirit of Judith Blau and Dana Cuff would reveal. The competences required are related to business management, financial management, people management, design management, leadership and governance.

Managing Projects

Writing in 1964, Brunton et al. noted that there was enough knowledge about managing projects, and so they concentrated on office management. However, since the 1960s, it is construction project management literature that has proliferated, with numerous books and research articles dedicated to the management of construction projects. While much can be learned from this body of knowledge, it is the design management literature that is most pertinent to architects. This literature is a sub-set of the architectural design and management body of knowledge, and it is also in need of rigorous research and further development to encompass digital architecture. Projects fuel the architectural business, and the effective management of individual projects will impact the profitability of the business. The way in which individual projects are managed will also impact the day-to-day experiences of individuals. Effective management usually translates to a happy workforce and considered workload planning. Ineffective management is usually evident in staff working long hours and deadlines being missed (Emmitt 2024). This has been highlighted as an ongoing issue (Sang et al. 2009), and it could be argued that the application of architectural management, together with digital tools, could go a long way to resolving the challenges that too many employees face. Digital tools such as BIM and digital twins are important for facilitating co-ordination of information and the smooth flow of work. Opportunities for further research relate to how the design project portfolio is managed in a digital milieu, and how this translates to better effectiveness of service delivery and better realisation of design value. The competences required are related to project management, information management, commercial management, design management, process management and people management.

Managing Stakeholders

Alharbi (2013) provided an original contribution to the architectural management literature by separating out the need to manage stakeholders as an activity that is often taken for granted. By identifying stakeholder management as a separate component of architectural management, it becomes necessary to confront it, both in the literature and in daily practice. This entails an understanding of how people behave at boundaries or interfaces within projects, and especially within mixed-mode digital project teams and diverse communities. The stakeholders involved with the business tend to be relatively consistent, whereas the stakeholders involved in projects will differ for every project. Similarly, engagement with the local community will be related to the geographical location of the architectural business and the location of the project. Opportunities for further research relate to how stakeholders engage with architectural businesses, and vice versa, in a digital marketplace. This also relates to the type of digital language adopted by the project team and stakeholders to facilitate effective working and engagement with communities. Evidence from projects has shown how digital tools have been used to consult and communicate with stakeholders in geographically difficult to reach projects. Moreover, it has been demonstrated how digital tools such as BIM and augmented reality can be used to collect site data and involve local people in the co-design and co-creation of buildings, helping users to effectively communicate their needs and preferences. However, there are concerns that digitalisation may be the province of the developed world and can exclude some communities. The competences required relate to people management, interface management, interpersonal communication and ethics in the widest sense of the term.

Managing Learning

Professionals have an obligation to manage learning throughout their careers. Internationally, there is variation in the way lifelong learning is managed, with some professional institutions setting targets and others suggesting guidance. This should be supported using reflective and reflexive practice (Schön 1983). The uptake of remote learning and the provision of continuing professional development (CPD) via online platforms have made lifelong learning more accessible, although deciding what is most important for individuals and their employer's business can be perplexing. In the context of this book, the learning need is linked to the opportunities offered by a wide variety of digital tools and, by extension, the link to management issues. The education of architects, especially the amount of management that should be and can be included in the curriculum of an architectural programme, has often been debated within the CIBW096 community (Daws and Beacock 2009; Svetoft 2009), and education is central to the CIBW096 Research Roadmap. The digitalisation of the construction sector and the digitalisation of architectural practice must be reflected in the curriculum for students of architecture, and not just assume that they will know what to do. Again, there are direct and indirect links to these opportunities, both in terms of general education and openings for specialisation. Openings for further research relate to how digital technologies can enhance learning for architectural students, for example, through better integration of architectural knowledge using machine learning. The competences required relate to reflective practice, knowledge integration, change management and research.

The chapters of this book are introduced under their section headings.

Technologies

Future challenges may be associated with better integration of digital tools and training for staff. Design skills are still essential, and fundamental to generating high-quality buildings, but so too is the ability to see market opportunities and manage tools such as AI and machine learning. They can contribute to the better prediction of, for example, building performance as identified in Chapter 2. BIM, digital twins and parametric and generative designs are important for effective design realisation and effective management, facilitating co-ordination of data and the smooth flow of work. In Chapter 3, it is shown how they can organise information for design, construction and use of buildings as well as better community engagement. Big data, smart buildings and the Internet of Things allow architects to be better engaged with buildings in use, which informs future designs and the creation of buildings that perform as intended. Similarly, blockchain technology provides opportunities for conceiving, using and re-using buildings in the context of a circular economy and addresses the United Nations' (2023) sustainable development goals

The Human Context of Digital Work

Stakeholder management demands an understanding of people, including behaviour, communication, knowledge sharing and integration, trust and governance, which are explored in Chapter 4. The way digital tools impact creativity in architectural projects is attracting attention from researchers and explored further in this book, especially in Chapter 5, which also analyses how different stages of the design process could be enacted. In addition, digital technologies provide the prospect for architectural businesses to be more inclusive places to work, populated by a diverse workforce and fuelled by equity in thought and talent. Diversity may also be reflected in more inclusive digital project teams. Issues of inclusion, equity and diversity are explored in Chapter 6, which implicitly highlights the importance of people in architectural design and management. Digitalisation can facilitate communication, collaboration and engagement with local communities in the widest sense of the term as shown in Chapter 7. Significant advances in digital tools are permitting the design and delivery of projects by diverse teams located in a variety of physical localities and within cyber space.

Emerging Concepts

BIM has contributed significantly to a greater degree of standardisation in process and product, as demonstrated in Chapter 8, with BIM standards contributing to the realisation of design value and to a more consistent level of service. Customisation is achieved through additional activities and plug-ins. Design is a non-linear process with iteration as a feature. This chapter investigates as to how it relates to a digitalised command sequence that rationalises the design process. The original objective of computer-aided design was to enable the human designer and computer to work together. It is this relationship between designer and design tools that is central to architecture, with the authors of Chapter 9 positing that architecture is essentially about people and the digital tools are a means to an end. A similar parallel could be drawn with architectural management tools and techniques, which are also a means to an end in realising architectural value. The authors propose a possible approach with AI, where the architect defines the scope and the machine provides suggested answers.

Managing Digital Projects

This section links to how the physical and virtual workplace is managed at a business level. It prompts guidance on how architects could organise digital projects and data-driven processes, as explored in Chapter 10. It also relates to how digital tools are changing procurement and contracts, and hence responsibilities and liabilities for design decisions, as in Chapter 11. It is noted in Chapter 12 that they allow designers to be creative, while also working collaboratively in physical and cyber environments, to ensure consistency of service and to create the best possible design value. Integrated project design and delivery can enhance collaboration and replace established adversarial relationships. The chapters in this book clearly set out the need for architectural businesses and architects to be conversant with available technologies and to keep abreast of the changing language of architecture. This is particularly expressed by the authors of Chapter 13. They also identify, both directly and indirectly, opportunities for architects to work in different ways and seize the openings offered by digitalisation.

Conclusion

It was at a CIBW096 conference held in Taiwan dedicated to future trends in architectural management (Tzeng et al. 2009) that a conference attendee stated that what architects needed is . . . *not more management, but better management*. It is a sentiment that resonated with this author and one that has often repeated over the years while teaching management to students of architecture. Unfortunately, the comment came from an anonymous voice in the audience, and so it has not been possible to give due credit for the insight. It is, however, sound advice and aligns with this author's own rationale of trying to simplify managerial processes and protocols to make them useful, and hence valuable, for creative people and creative businesses. It is easy to make something complicated and overly complex. It takes more time, effort and intellect to make systems easy to use and relevant to the tasks involved. This applies to the design of buildings just as much as it does to the design of the management systems, tools and protocols used in daily architectural practice. It also applies to the teaching of management within architectural programmes.

Taking the need for better management forward, the introduction and widespread adoption of digital tools by architectural businesses provides the opportunity to streamline many of the mundane administrative tasks that distract designers from the more exciting tasks associated with designing buildings and realising design value. From the overview of the four components of architectural management, it is evident that there is a need for competent people, clear and consistent leadership, effective management of information and knowledge and application of appropriate digital technologies. This influences the culture of the architectural business and the projects that feed it. Indeed, as highlighted and implied in the chapters that comprise this book, it is the creative application of design processes that include digital tools in conjunction with management that will bring about a transformation in culture and enable architectural businesses to collaborate and compete in a dynamic marketplace for services.

There is considerable synergy between the themes in the small body of architectural management literature and those within this book. The earlier work has primarily focused on management and its application to creative businesses and projects. While this has included people, the human context appears to need further enquiry and exploration to better address the competences and knowledge required for a digital age. Architects are not well versed in managerial thinking when they initially qualify, yet it is the leadership skills and the ability to bring about change that are so desperately needed in many architectural businesses. This reinforces the CIBW096 commitment to engage with architecture programmes around the world to raise awareness and promote architectural design and management principles. The challenge for architects is that the literature available is modest in volume and research findings into the daily work of the architect are scarce. The professional journals tend to run articles based on experience, which is helpful, but which is rarely informed by research. The latter is required to explore the human condition in a digital architectural milieu to inform architectural practice about how best to manage the people and technology interface. This brings the narrative back to the CIBW096 Research Roadmap (Giddings and Melhado 2019), which highlights the importance of CIBW096 as a facilitator of research. Stimulating more research into architectural design and management and encouraging the application of research findings by architects in the workplace would be a positive development to the continued evolution of architecture and architects' contribution to a more appropriate, sustainable and ethical built environment. This book positions CIBW096 Architectural Design and Management at the forefront of architectural research, education and practice.

References

Alharbi, M.A.S. (2013). *Architectural Management: a strategic framework to achieve competitiveness*. PhD Thesis. Loughborough University.

Allinson, K. (1993). *The Wild Card of Design: A Perspective on Architecture in a Project Management Environment*. Oxford: Butterworth Architecture.

Blau, J. (1984). *Architects and Firms: A Sociological Perspective on Architectural Practice*. Cambridge, MA: The MIT Press.

Bondare, S., Volkova, T., Bondars, E. and Barvika, S. (2017). Business challenges in the architecture industry. *Architecture and Urban Planning*, 13(1), 118–123.

Brunton, J., Baden Hellard, R. and Boobyer, E.H. (1964). *Management Applied to Architectural Practice*. London: George Godwin.

Cuff, D. (1991). *Architecture: The Story of Practice*. Cambridge, MA: The MIT Press.

Daws, M. and Beacock, P. (2009). Incorporating management into an undergraduate architectural design programme. In S. Emmitt, M. Prins and A. den Otter (eds.) *Architectural Management: International Research & Practice*. Chichester: Wiley-Blackwell. 297–312.

den Otter, A., Emmitt, S. and Achammer, C. (eds.) (2011). *Architectural Management in the Digital Arena*. Proceedings of CIBW096, Vienna Technical University, Austria.

Emmitt, S. (1999). *Architectural Management in Practice: A Competitive Approach*. Harlow: Longman.

Emmitt, S. (2024). *Design Management for Architects* (Third Edition). Abingdon: Routledge.

Emmitt, S. and Alharbi, M. (2018). *Handbook for the Architectural Manager*. Chichester: Wiley-Blackwell.

Emmitt, S., Prins, M. and den Otter, A. (eds.) (2009). *Architectural Management: International Research & Practice*. Chichester: Wiley-Blackwell.

Freling, W.V.J. (1995). Architectural management: a profession or specialist interest? *International Journal of Architectural Management, Practice and Research*, 9, 11–20.

Giddings, B. and Melhado, S.B. (2019). *Architectural Design and Management Research Roadmap*. CIB Publication 416. Rotterdam: CIB.

Gray, C., Hughes, W. and Bennett, J. (1994). *The Successful Management of Design: A Handbook of Building Design Management*. UK: University of Reading.

Maister, D.H. (1993). *Managing the Professional Service Firm*. Glencoe, IL: The Free Press.

Nicholson, M.P. (1992). *Architectural Management*. London: E & FN Spon.

Nicholson, M.P. (1995). Architectural management: towards a definition. *International Journal of Architectural Management, Practice and Research*, 9, 1–10.

Nicholson, M.P. and Prins, M. (1993). Architectural Management, Practice and Research. *Proceedings of the first CIBW96 Workshop*. Eindhoven University of Technology.

Nikseresht, M., Mozaffar, F., Hosseini, S. B. and Sedghpour, B.S. (2022). Management in architectural design: a reflection on the profession and education. *Industry and Higher Education*, 36(5), 680–685.

RIBA (1962). *The Architect and His Office*. London: Royal Institute of British Architects.

Sang, K.J.C., Ison, S.G. and Dainty, A.R.J. (2009). The job satisfaction of UK architects and relationships with work-life balance and turnover intentions. *Engineering Construction and Architectural Management*, 16(3), 288–300.

Schön, D.A. (1983). *The Reflective Practitioner*. New York: Basic Books.

Sharp, D. (1986). *The Business of Architecture*. London: Collins.

Svetoft, I. (2009). The architect's role. In S. Emmitt, M. Prins and A. den Otter (eds.) *Architectural Management: International Research & Practice*. Chichester: Wiley-Blackwell. 284–296.

Symes, M., Eley, J. and Seidel, A.D. (1995). *Architects and Their Practices: A Changing Profession*. Oxford: Butterworth Architecture.

Tzeng, C., Emmitt, S. and Prins, M. (eds.) (2009). Future Trends in Architectural Management. *Proceedings of the International Symposium CIBW096*. November. Taiwan.

United Nations. (2023). *Tier Classification for Global SDG Indicators*. Report. New York: UN. https://unstats.un.org/sdgs/files/Tier%20Classification%20of%20SDG%20Indicators_13%20Dec%202023_web.pdf.

Part II

Technologies

2

The Digital Toolkit

Ania Khodabakhshian[1], Taija Puolitaival[2], and Linda Kestle[3]

[1] *UCLA, Los Angeles, USA*
[2] *Tampere University, Finland*
[3] *Unitec Institute of Technology, Auckland, New Zealand*

Introduction

After the three industrial revolutions of mechanisation, electrification and automation, the architecture, engineering and construction industry has been experiencing a digital transformation; thanks to a new revolution, as a result of which abundance of digital and internet-based tools are being used to support decision-making through the design and management of architectural projects (Boje et al. 2020). This phenomenon is happening for three main reasons. The first is Big Data, where an immense volume of heterogeneous information from diverse disciplines is being produced, generated and documented in the industry in various formats of text, images, etc. throughout the entire life cycle of buildings (Yan et al. 2020). The second is Digital Technologies, including Building Information Modelling (BIM), wireless networks for real-time data collection, robots, laser scanners and sensors (Štefanič and Stankovski 2019). Third, is Artificial Intelligence, which can process large amounts of data, extract patterns and valuable insights, and model, predict, automate and optimise processes. Given the rapid advancements in digital technologies, companies can now deal with petabytes of data, which can offer novel opportunities to assess and improve their efficiency and make data-driven decisions with the help of statistical and algorithmic models. Data is the key driver of automation of manual and repetitive processes, and various technologies are being developed in accordance with each input data type.

One of the most common digital tools used in the Construction Industry is BIM. It is an example of the evolution from manual to simple, digital and gradually more complex tools. BIM evolved from Computer-Aided Design or more precisely from Computer-Aided Drafting. CAD was introduced in the 1970s to replace drawing boards. It enabled rapid replication and, therefore, faster production of drawings. However, a two-dimensional CAD drawing often just appeared as a complex collection of lines, symbols and text

Architectural Design and Management in the Digital Age: International Perspectives, First Edition.
Edited by Bob Giddings, Silvio Melhado, and Jenni Barrett.

information; and co-ordination between the drawings needed to be completed manually. Every time a line was changed, all the related lines and drawings needed to be manually modified. Information sharing and co-ordination between different design disciplines were also poor. The drawings contained only visual representation of the design, and the rest of the information was in separate documents or just in the head of the designer (Ingram 2020). Even though two-dimensional CAD gradually developed into a three-dimensional version, allowing better visualisation of the building, the information sharing and co-ordination issues remained. There were numerous development paths, ultimately leading to BIM (Ingram 2020). Although the term and concept of BIM became known to the industry in early 2000, many early developments contributed to it. In 1962, Douglas C. Englebart painted a vision of object-based design, parametric manipulation and relational database. In 1975, Charles M. Eastman provided a more developed and concrete concept of a Building Description System, with three-dimensional objects corresponding to real building elements (Eastman et al. 2011), and Frank Gehry's use of CATIA, an aircraft design software, in the early 1990s, is probably the most well-known early example of a computer-aided three-dimensional design in architecture (Briscoe 2015).

Alongside BIM, smart machinery like robots and 3D printers are now used for design prototype production, comparison of alternatives and building material optimisation, enabling the design and construction to take place in some of the most difficult locations. Networks of interconnected physical devices for image and data acquisition, such as cameras, drones, sensors, radio frequency identification devices and laser scanners, help gather real-time and as-built data from projects to react proactively with the requirements or problems in the design iterations. The integration of various novel technologies such as Augmented Reality, Virtual Reality and Mixed Reality with BIM creates productive interactions between the physical and cyber worlds. Blockchain is another robust shared global infrastructure for simplifying and securing transactions amongst parties and for purposes like digital signage and management of contracts. It is a decentralised and distributed digital ledger used to record transactions across multiple computers in a way that ensures each transaction can only be modified or edited if all subsequent blocks in the chain are altered. Artificial Intelligence, a significant branch of computer science, is the new frontier, already proven to be highly effective in other sectors like finance and medicine. It can be used to develop computer systems and smart machines capable of performing tasks that normally require human intelligence, reasoning, analysis, knowledge, planning and problem-solving. As a result, manually performed processes can become automated, optimised, cost-effective, less time-consuming and self-modifying without human intervention. In addition, it can learn from massive, complex and dynamic data from similar previous projects. Artificial Intelligence can perform pattern recognition, prediction, risk mitigation, process digitalisation and decision-making optimisation (Madanayake and Egbu 2019). The integration of Artificial Intelligence with Digital Twin models like BIM in the form of generative design alternatives or automated rule checking can automate and optimise the design process with respect to specific design requirements and regulations. A Digital Twin is a digital representation of a physical entity, i.e. a virtual model of a building or infrastructure, including not only its architectural and structural elements but also its systems, processes and even its performance throughout the life cycle. Architectural design and management benefits from several digital tools and processes. Figure 2.1 is a keyword

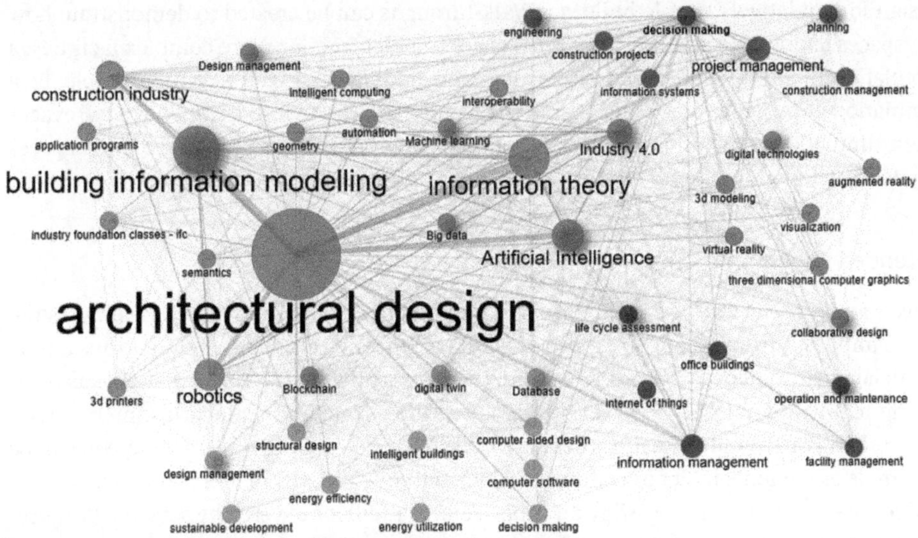

Figure 2.1 Architectural design and management and related digital tools and processes.

co-occurrence diagram based on a literature search conducted to delineate the connections between these digital tools. This figure helps to identify the main digital tools studied in literature, their application scope, the project phase they are used in and their interrelations with each other.

Visualisation, Simulation and Modelling

Building Information Modelling

The most common uses of BIM are communication, collaboration and co-ordination between design disciplines and the whole project team. The nature of BIM as a three-dimensional model lends itself well to the visualisation of the design. The added information improves communication and enables various simulations to be run. In addition to the visualisation and simulation, various analyses, including sustainability-related assessments and three-dimensional co-ordination can be derived. Visualisation during the design review process is vital to assess how well the design is meeting the requirements. Applications, such as Revit and ArchiCAD, can be used to visualise the design. However, there are also specific tools for this purpose, including many freeware, some of which are desktop applications and some web based. Freeware is available but only suitable when there is no need to change the model while viewing it and is also better for non-experienced users as the interface is usually easy to use with only a limited selection of functions. An essential aspect of the design review process is design co-ordination, and there are automatic processes for it. Visual co-ordination is important as the starting point and at the early stages of the design when there is just one BIM model to review. A visual assessment using the BIM model is usually adequate to assess the usability and functionality of the design for a relatively simple building. Walk-throughs can be created to demonstrate how the spaces look inside and how different spaces interact. Yet, for more complex facilities, a simulation might be more informative and value-adding. This could, as an example, be a simulation of people's movement within the facility, such as an airport or a hospital, evacuation simulations or simulating work functions in a production facility to see if the spaces support the functions.

Extended Reality

This is an umbrella term for Virtual Reality, Augmented Reality, and Mixed Reality. As the name implies, Virtual Reality includes only fully virtual objects and therefore forms a fully artificial environment. Artificial Reality and Mixed Reality both have virtual objects overlaying real-world objects. However, in the latter, the virtual and real worlds interact with each other. For Extended Reality, there are multiple technology options. The screen can be as simple as a phone, tablet or computer screen; although there are more complex systems called caves – rooms with possibly all walls, ceiling and floor covered by screens. Purpose-built glasses are another option. Different sensor technologies such as motion detection, gaze tracking and speech recognition are used for navigating the space and interacting with and controlling elements in the spaces. These sensors are attached to wearables such as glasses, gloves or bracelets. This is very much an extension of BIM, where it is the virtual environment or the origin of individual virtual objects. It can be used for design analysis and reviews to enhance visualisation while simultaneously communicating the design (Suleiman and Ayer 2019). Artificial Reality and mixed reality are helpful with renovation, retrofitting and extension projects, where the connection between the new and the existing elements is critical. Extended Reality therefore offers an improved spatial understanding of the functionality of the design to support decision-making (Suleiman and Ayer 2019).

Artificial Intelligence

Building data and codes are becoming more accessible due to the advancements in acquisition gadgets and mining techniques. In particular, Machine Learning algorithms have a critical role in generating and evaluating design models before construction starts (Hong et al. 2020). Data-driven and Artificial Intelligence-based applications for project visualisation and simulation are mainly Parametric Design, Generative Design, Design Evaluation, Cloud Computing and Computer Vision. However, data gathering technologies are inseparable components of these processes, which provide the raw input data to be processed and simulated. Data can be connected and exchanged with remotely located devices and systems over the internet, which significantly lowers the data gathering, processing and management costs (Štefanič and Stankovski 2019). Machine Learning-based Parametric Design is used to reproduce the hard-coded rules and parameters and discover the underlying phenomena behind them. For instance, Principal Component Analysis can help understand the design variables with the highest effect and variance so that more focus is put on them, resulting in an optimised design process. Generative Design is an iterative process that leverages AI and Machine Learning to generate several design solutions and outputs meeting certain constraints, objectives and rules. The designer selects the initial constraints and values, and has the final decision on the design choice. Therefore, generative design has a high level of human–computer interaction. Machine Learning-based Design Evaluation fosters a design decision-making process with a focus on space utilisation, safety, energy efficiency and occupant well-being. This process aims to predict the quality and performance of the designed space (Hong et al. 2020).

Analysis

Building Information Modelling

The information contained in BIM, enables not just communication and simulation, but also analyses. Which particular analyses can be done depends on what information has been added to the model, and how detailed and correct the model geometry and information are. As modelling building elements and adding information is resource-consuming, the extent of modelling needs to be carefully considered at the start of the project. The modelling work should be weighed against the benefits. These can be either internal for the design practice or external, i.e. benefits to other project parties or both. As the essence of BIM is data management throughout the building life cycle, the benefits and cost of modelling should be discussed with the other project parties and shared as appropriate. The number and accuracy of the possible analyses increase the further the design develops, as the more detailed the geometry and the information become available. Applications are emerging, which can be used as early as the concept design stage to support decision-making. Automatic three-dimensional co-ordination of design within and across disciplines is one of the most popular uses of BIM. In a large project, it is common to share the design authoring responsibility inside an architectural practice. This can create separate models, so there is a need to ensure that there are no geometrical clashes or object duplications in the interfaces of the models. Similarly, and even more important, the different discipline models need to be co-ordinated with each other. This is done not just to identify and solve hard clashes where two physical items intersect but also to identify so-called soft clashes, where two or more items are in too close a proximity to each other for installation or maintenance purposes. The most common software tools for three-dimensional co-ordination are Solibri and Navisworks Manage. Occasionally some freeware is involved, such as the former Tekla BIMSight. Solibri can also be used for other automatic checking of the model, as the operator is allowed to modify the existing checking rules and develop new ones. This allows designers to check the design against the building code and analyse issues such as the functionality or how well the model matches the initial spatial program. Modifying and creating new rules is time-consuming for one project. However, if the same rules can be used more than once, the time investment can be justified.

Building sustainability life-cycle assessment can also benefit from the use of BIM. The process requires information about the building material and the building energy consumption for buildings. There are numerous software tools for life-cycle assessment. Most tools can be used for all types of buildings, but geographical coverage in energy consumption databases varies. Some tools are simple web-based forms, where the information needs to be manually extracted from the design documentation and entered onto the form.

Similar to building sustainability assessment, when calculated manually, life-cycle assessment is most often completed too late in the project to inform decision-making. However, many of these tools have the possibility for automatic quantity take-off from the BIM model. Some tools such as Athena Impact Estimator (free, North America) and Greencalc (free, Europe) can even be linked both to the design and energy modelling tools to accommodate design changes and, therefore, can be applied earlier in the project when design decisions are being made. A comprehensive review of the features and user-friendliness of 16 tools is available by Karunaratne and Dharmarathna (2022). As life-cycle assessment relies heavily on material information, it is evident that it cannot be used at the early design stage when this information is not yet available. Energy performance assessment evaluates a building's energy consumption and can also include greenhouse gas emissions. With the introduction of BIM, the analyses have become easier to complete as the objects in the BIM model can include information for the analysis, and a more general understanding of energy consumption and its factors is adequate to perform the analysis. There are various combinations of design-authoring tools such as Revit and ArchiCAD, energy assessment tools such as Green Building Studio and DesignBuild and simulation engines such as EnergyPlus and e-Quest for energy performance assessment. It is possible to perform a simple assessment at the concept design stage based on the shape and size of the building, building orientation and building topology. At the preliminary design stage, energy use, lighting and daylighting, sun and shadow analysis, heating, air flow and ventilation analyses can be affected and completed, either visually from the model or via a simulation in the case of the air and heat flow analysis. These can be re-calculated at the detailed design stage.

There are other analyses, such as cost estimation and life-cycle cost assessment that can be undertaken using BIM. Although it adds value to cost estimation, it is not a specific activity for the architectural discipline rather a task for quantity surveyors. However, it needs to be noted that it is essential for the design author and the cost estimator to discuss how the building elements are to be modelled and what information to include in the model for the quantity take-off to be error-free. Partly automated cost comparisons of design alternatives, in turn, allow the cost estimations to be run more often. Life-cycle cost assessment evaluates the total cost of building ownership, including the initial investment, operation, facility management, renovations and disposal. This is valuable information for an owner-developer when considering different design options. Prototypes for life-cycle cost assessment tools, which use BIM and game engines to quickly and visually provide the cost when materials change, have been specifically developed for owner-developers (Omaran et al. 2019). The calculations and comparisons come with some challenges – fluctuations in energy prices, raw material shortages, new innovations and regulations, and long life cycles make it challenging to assess the cost. There are no direct tools to assess life-cycle cost as such. However, some of the tools for sustainability purposes described above can be used to assist with the calculations for operation, facilities management, renovation and disposal, and cost estimation tools can be employed too.

Artificial Intelligence

Various types of data are produced for projects, such as design data and drawings, schedule and cost baselines, contracts and project charters and enterprise resource plans. Various methods have been used for data analysis, such as Statistics, multi-attribute decision support systems such as Analytical Hierarchy Process, Data Mining, Predicting Analytics, Business Analytics, Knowledge Discovery from Data, Data Science and Analytics, Big Data and Process Mining. All of these methods have one aspect in common, their foundation in factual historical data (Bilal et al. 2016). They are used for text analytics, content-based analytics, structure-based analytics for cloud-based system framework for structured BIM data, audio analytics, visual-image and video analytics, social media analytics, geographic information system (GIS) analytics and predictive analytics (Madanayake and Egbu 2019). Statistical techniques like Decision Trees, Factor Analysis, and Monte Carlo Simulation have been employed to identify the causes of design and construction process delays and decision-making for litigation. However, when applied to large data sets from different sources, statistical algorithms need to be carefully designed with a special focus on computational cost reduction and real-time data streams. Another common challenge is poor data registration quality and the existence of irretrievable missing values. Therefore, when facing a limited database with missing values, deterministic models like Artificial Neural Networks cannot perform properly, and the companies should apply probabilistic models like Bayesian Networks instead. Fuzzy Logic integration with Bayesian Networks provides the opportunity to integrate the experts' subjective opinions into the computational model, which compensates for the data scarcity problem. In general, Fuzzy Logic can model nonlinear functions, tolerance of imprecise data, and blend with conventional control techniques to build on top of the experience of the experts (Mehmood et al. 2019). It is done by converting linguistic data gathered from questionnaires and interviews into crisp values, which mainly serves as an input for risk impact and probability analysis in projects and automates the manually conducted risk analysis process.

A more powerful method for automatic exploration and analysis of large data volumes is Data Mining, which benefits from Machine Learning and statistics to discover patterns and relationships in large databases with multi-attribute information. Machine Learning, as a highly applied sub-field of Artificial Intelligence, in practice, tends to enable computers to learn from previous data and generalise the rules and patterns to new cases for supervised and unsupervised learning (Hong et al. 2020). Supervised Learning is applied for issues with a known set of outcomes like regression models and classification problems. For example, predictions such as design alternative costs, litigation outcomes, integrated planning for optimal resource assignments, risk identification and analysis and automated project document classification. Supervised Learning is done by techniques such as Support Vector Machines, Latent Semantic Analysis, Genetic Algorithm, Decision Tree and Random Forest, Bayesian Network, Fuzzy Logic, Artificial Neural Network and Fault Tree Analysis. Unsupervised Learning is applied to issues with an unknown set of answers like clustering of buildings based on their energy performance metrics, developing retrofit strategies for building stock, and building image segmentation for further analysis using techniques like K-means clustering algorithm, Gaussian Mixture Model algorithm and Mean-Shift clustering algorithm. Alongside problem-based analysis, Artificial Intelligence can play a crucial role towards integrated, smart and sustainable design projects. An ideal maturity level is to make the whole design, construction and operation processes intertwined and collaborative, with a long-term perspective on added values and consequences of each decision and design choice. For example, benefiting from environmental data captured by Arduino and sensors in a specific zone, it is possible to implement accurate design solutions in accordance with the environmental conditions at the micro-level. Approaches of this kind can be implemented for resource and waste management, water conservation, indoor natural lighting and air quality and green building principles application, all of which lead to a more sustainable built environment. BIM contains significant financial information on buildings through the entire life cycle and can feed this cost information algorithms for automated retrieval and optimisation of life-cycle cost, i.e. providing cost-optimal replacement of building components, deciding the economic building design schemes and predicting additional costs at BIM's different levels of development (Zhang et al. 2022).

Automation, Optimisation and Prediction

Significant amounts of structured and unstructured data are produced in projects every day. The automatic classification and resultant learning facilitate management and decision-making. Structured data, such as sensor data, are easier to handle, and many information platforms like BIM and GIS are being developed to use it. More than 80% of produced data is unstructured, e.g. images and textual data, including design specifications, BIM logs and project contracts. Artificial Intelligence-based Text Mining techniques including Natural Language Processing are beneficial to mining patterns, trends and rules from the unstructured text data in a cost-effective manner. It functions by transforming unstructured text into a structured format, with many tokens and terms using preprocessing techniques and weighting them, based on their importance within the documents. The main applications of Natural Language Processing in projects are (i) filtering and organising documents, e.g. finding accident causes from reports or grouping drawings from different disciplines, (ii) developing expert systems, e.g. integrating expert knowledge and providing knowledge-based answers to design problems, (iii) automated compliance checking, e.g. automatic comparison between as-planned and as-built components and detecting non-compliances (Wu et al. 2022). Artificial Intelligence and Machine Learning algorithms have beneficial applications during preprocessing and design phases for automatically checking building designs and BIM models for code compliance, clash detection and quality assurance, saving time and cost (Madanayake and Egbu 2019). These techniques are usually adopted to predict or assess the design objectives of the built environment concerning different design alternatives in the early decision-making stage of the design process. Artificial Intelligence can bring multiple benefits such as energy-saving, waste reduction, and cost-saving to the building design process. Parametric Design and Machine Learning are examples of integrated design processes. Decision Trees and Support Vector Machines have been applied to predict the performance of natural indoor daylight, and BIM has served as a design decision support system for automated rule checking and customised information retrieval in previous research. Another important application of Artificial Information techniques, especially Artificial Neural Networks and Decision Trees, is cost estimation during the design stage, which is a vital activity for the successful implementation of a construction project and keeping the clients informed as the design process progresses.

Materials performance prediction is another added value that it brings to the design stage, using techniques like Support Vector Machines, multiple regression, and Artificial Neural Networks. Prediction using Artificial Intelligence is mainly achieved in two forms: (i) Supervised Learning for classification and regressions problems, where classification is automatic, assigning a class label based on previous samples. For instance, it can be used to predict the post-retrofit energy label of a building based on the proposed retrofit strategy. Regression is the prediction of future values and trends, for instance, predicting construction tender prices and material price fluctuations, which directly influence the design choices. (ii) Unsupervised Learning for Clustering data points without clear labels into groups with similar characteristics. For instance, clustering building envelopes with similar thermal transmittance (Pan and Zhang 2021). Optimisation seeks and delivers more sustainable project plans and decisions under the optimal trade-offs of time, cost and quality, based on a set of criteria and constraints, applicable locally and globally. Optimisation algorithms like Genetic Algorithms and evolutionary programming can be used for the design, construction and operation phases. Evolutionary computation techniques including procedural algorithms, heuristic search-based methods, genetic algorithms, fuzzy logic and artificial neural networks are algorithms for space planning, based on the findings of relationship and adjacency diagrams, developed for visualisation (Dutta and Sarthak 2011).

In the design phase, optimisation algorithms are used to find economic architectural forms and select optimum structural design solutions for a series of design constraints. It is mainly achieved by evolutionary computation, which is based on replication of natural structures through simulation of evolution with respect to a set of parametric relationships and rules. These generate systems that adapt to their environment in a similar manner to natural selection to make the design more flexible and efficient. It is noteworthy that aside from optimisation, there is often a focus on reducing design costs for creating innovative and high-performing forms. Additionally, it is useful for developing cost-optimal and zero energy buildings with less computational effort. Project planning, resource allocation and reduction in energy consumption are other applications. Another area associated with optimisation is Risk Mitigation and Management. Adopting Artificial Intelligence makes it possible to identify, assess, prioritise and predict risks related to design and construction phases in an automated, unbiased and real-time manner. The most common methods for Risk Management are probabilistic models like Bayesian Networks, Fuzzy Theory and Deep Learning Algorithms such as Neural Networks. For instance, it is possible to capture interdependencies of causes and accidents; measure the probability of failure occurrence, early troubleshoot and prioritise potential risks caused by a design solution and determine pre-emptive actions to keep projects on time and budget. Currently, the most used design optimisation techniques are Artificial Neural Networks, Fractals, Swarm Intelligence and Cellular Automata. The main applications are design exploration using generative or modular design, morphogenesis using parametric and cellular systems and building shape and components from optimisation (Castro Pena et al. 2021).

Automation using Artificial Intelligence can address the bias and confounding nature of conventional, time-consuming and manual methods. It can be integrated into project management software to facilitate automatic data analysis and decision-making based on historical data and tacit knowledge hidden in large databases. Automation examples are

Additive Manufacturing techniques like 3D printers for prototyping and comparing design solutions, on-site monitoring using drones and sensors to record data and create images about the construction status and progress. Smart machinery such as robots to potentially take the place of on-site workers can assist efficiency, productivity and even profitability. Digital fabrication is another example of data-driven automation that might seem like a double-edged sword for the design discipline. On one hand, it enables a direct line from design to fabrication and assembly of complex and unique forms using additive manufacturing. On the other hand, it implies standardisation and modularisation of design by automated prefabrication, although architectural creativity may still be used to design modules and their combinations in a way that enables the production of unique and diverse building designs. BIM has a role in digital fabrication, providing the framework for detailed fabrication design. BIM can support the automation of prefabrication and even automation of construction and assembly on site.

Although manufacturing-specific software also exists, BIM software such as Revit and even SketchUp can be used to run 3D printers and AutoCAD for CNC routers and other cutting tools. Another area of digital tool application through the design phase is Prevention through Design in four main areas: (i) knowledge-based systems, (ii) automatic rule checking, (iii) hazard visualisation and (iv) safety training for designers – all of which have a significant effect on reducing safety risks involved in the construction and operation phases. Knowledge-based systems provide fundamental information to enhance the Prevention through Design competence of designers through collections, filtering, comparing and analysing information from experts, facts and observations. It is possible to link Health and Safety risk-related knowledge to relevant building objects in the BIM environment during the design phase. Management risk, financial risk and legal risk are other important aspects that can be added to the BIM model for future automated reporting and quantity take-offs. Automatic Rule Checking facilitates the evaluation of the design solution with regulations and codes of practice through encoding rules and criteria into the virtual building model. Usually, the BIM platform is used to implement rule checking and reporting results as pass, fail or warning. Automatic Rule Checking can significantly reduce rework and possible clashes with consequent time and cost overruns. For Hazard Visualisation, various technologies such as BIM, Artificial Reality and Virtual Reality offer a three-dimensional environment for visualising high-risk tests, temporary structures, safety programs and safety and productivity of site workers for better identification and management of hazardous areas. Safety Training is spatially important for designers to understand and know how to adopt digital technologies for enhancing Prevention through Design processes (Farghaly et al. 2021). Although parts of the design can already be automated using parametric and generative design, and even manufacturing of repetitive and modular elements can be automated, the industry is still quite distant from Artificial Intelligence and robotics replacing humans entirely in architectural design and management. The machines still lack human abilities to create, innovate, negotiate, empathise, and understand aesthetics. A fruitful direction for the development would be intelligent collaboration, where the machine would focus on complex calculations and data analysis, and the human on emotional intelligence and aesthetical strengths in the project (Lin and Xu 2022).

Conclusion

Implementation Process

In entering the digital era, conventional job roles, business structures and processes are changing, and new sets of skills, knowledge and training are required to use the full potential of digital tools and be able to compete in the industry. Adopting new processes and technology needs to be seen as an investment as it comes with an initial cost, and the benefits are not immediate. Computer hardware and software investments can be arranged traditionally by owning hardware and individual or floating licenses, through software-as-a-service or more holistic platform-as-a-service solutions via the cloud. While full software applications such as Revit and Solibri are expensive, there are providers who offer viewing applications for free. Although the freeware is not suitable for design authoring, simulation or automated analysis, they provide a simple interface for visualisation and communication. Additional hardware investments might include extended reality wearables, various digital cameras and drones, depending on what is perceived as key in supporting the processes. There is a lack of skilled professionals and domain experts who can properly and continuously use digital tools and integrate them into a company's processes. Therefore, special training and continuous learning are required at the enterprise level. This includes not just training courses but, more importantly, learning at work with extra time allowance for project work to adopt the new tools and processes. The large software vendors and their partners offer training for basic software processes. However, for more specific software applications and especially for the overall project process, there is very limited, if any, training available. Professional institutes offer some process training for their members. Tertiary institutes in many countries have adopted new processes such as BIM as part of their curricula, but it takes time before these graduates are the majority in the industry. At the same time, the processes and digital tools keep developing, creating a continual need for professionals to update their competencies. However, by continuous learning and integration of the new tools and processes in projects, the long-term benefits and cost savings will be much greater than the short-term costs of technology implementation.

Integration

In the Big Data era, and with collective intelligence sharing of individuals through the technological environment, it is possible to create a common platform such Digital Twins, benefiting from BIM, Artificial Intelligence and Cloud computing technologies to document, measure and capture data for collaborative analysis and decision-making. Valuable insights can be extracted from structured or unstructured data and focus on data accessibility, communication and management. This results in a competitive advantage for companies over their business rivals. Therefore, Big Data plays the convergence role of people, place and technology and helps improve understanding of common data environments (Madanayake and Egbu 2019). Digital Twins is a broader context of digital and data-driven models like BIM and aims to improve the existing design and construction processes and models using interoperable formats to reflect the construction of physical assets in different phases of a project virtually (Boje et al. 2020). BIM modelling is the starting point for semantically rich three-dimensional knowledge and data entry. Web protocols guarantee data linking and interoperability, and Artificial Intelligence will facilitate the simulation, prediction and optimisation for applications in energy efficiency, facility management and smart cities. For cyber–physical integrated modelling, the real-time services data are transformed into optimised service data to demonstrate the gap between the current and ideal states of project performance. In addition, it draws an efficient road map to reach a more productive and successful stance in projects. Examples would be real life-cycle data such as supply chain information, site monitoring data to establish scheduled progress and site sensing or weather data, which are systematically registered in the BIM model. These then serve as input for the algorithms to produce optimisation data, for example, prediction data – cost forecasts, simulation and clash detection data and updated building data for the BIM model and project documents (Tao et al. 2018).

Future Directions

Future directions and applications of digital tools could be as follows: (i) Smart Robotics for carrying image and video acquisition systems and conducting automatic on-site works; (ii) Cloud Virtual and Augmented Reality, and its integration with BIM, for information visualisation and interaction; (iii) Artificial Intelligence as the incorporation of techniques into the infrastructure for more effective operation and data analysis; (iv) 4D printing for enabling 3D-printed objects to change their shape and behaviour over time, and respond to external factors like heat and light; (v) Blockchain as a robust shared global infrastructure utilised for simplifying and securing transactions amongst the parties involved, and its integration with BIM, which can enhance safety and operational performance in the built environment, and data sharing security amongst stakeholders and (vi) Digital Twins. This future directional forecasting may soon become the new normal in architectural design and management academic research and practice, to inform strategic decisions about using digital technologies in architectural projects.

References

Bilal, M., Oyedele, L.O., Akinade, O.O., Ajayi, S.O., Alaka, H.A., Owolabi, H.A., Qadir, J., Pasha, M. and Bello, S.A. (2016). Big data architecture for construction waste analytics (CWA): a conceptual framework. *Journal of Building Engineering*, 6, 144–156. https://doi.org/10.1016/j.jobe.2016.03.002.

Boje, C., Guerriero, A., Kubicki, S. and Rezgui, Y. (2020). Towards a semantic construction digital twin: directions for future research. *Automation in Construction*, 114(January), 103179. https://doi.org/10.1016/j.autcon.2020.103179.

Briscoe, D. (2015). *Beyond BIM: Architecture Information Modelling*. London: Routledge.

Castro Pena, M.L., Carballal, A., Rodríguez-Fernández, N., Santos, I. and Romero, J. (2021). Artificial intelligence applied to conceptual design. A review of its use in architecture. *Automation in Construction*, 124. https://doi.org/10.1016/j.autcon.2021.103550.

Dutta, K. and Sarthak, S. (2011). Architectural space planning using evolutionary computing approaches: a review. *Artificial Intelligence Review,* 36, 311–321. https://doi.org/10.1007/s10462-011-9217-y.

Eastman, C.M., Teicholz, P., Sacks, R. and Liston, K. (2011). *BIM Handbook: A Guide to Building Information Modeling for Owners, Managers, Designers, Engineers and Contractors* (Second Edition). Hoboken: John Wiley & Sons.

Farghaly, K., Collinge, W., Mosleh, M.H., Manu, P. and Cheung, C.M. (2021). Digital information technologies for prevention through design (PtD): a literature review and directions for future research. *Construction Innovation,* 22(4), 1036–1058. https://doi.org/10.1108/CI-02-2021-0027.

Hong, T., Wang, Z., Luo, X. and Zhang, W. (2020). State-of-the-art on research and applications of machine learning in the building life cycle. *Energy and Buildings,* 212, 109831. https://doi.org/10.1016/j.enbuild.2020.109831.

Ingram, J. (2020). *Understanding BIM: The Past, Present and Future.* Abingdon, UK: Routledge.

Karunaratne, S. and Dharmarathna, D. (2022). A review of comprehensiveness, user-friendliness, and contribution for sustainable design of whole building environmental life cycle assessment software tools. *Building and Environment,* 212, 108784. https://doi.org/10.1016/j.buildenv.2022.108784.

Lin, C.Y. and Xu, N. (2022). Extended TAM model to explore the factors that affect intention to use AI robotic architects for architectural design. *Technology Analysis & Strategic Management,* 34(3), 349–362. https://doi.org/10.1080/09537325.2021.1900808.

Madanayake, U.H. and Egbu, C. (2019). Critical analysis for big data studies in construction: significant gaps in knowledge. *Built Environment Project and Asset Management,* 9(4), 530–547. https://doi.org/10.1108/BEPAM-04-2018-0074.

Mehmood, M.U., Chun, D., Han, H., Jeon, G. and Chen, K. (2019). A review of the applications of artificial intelligence and big data to buildings for energy-efficiency and a comfortable indoor living environment. *Energy and Buildings,* 202, 109383. https://doi.org/10.1016/j.enbuild.2019.109383.

Omaran, S.M., Alghamdi, A.A., Alharishawi, S.C. and Hains, D.B. (2019). Integrating BIM and game engine for simulation interactive life cycle analysis visualization. In Y.K. Cho, F. Leite, A. Behzadan and C. Wang (eds.) *Computing in Civil Engineering 2019: Visualization, Information Modeling, and Simulation.* 120–128. American Society of Civil Engineers. https://doi.org/10.1061/9780784482421.016.

Pan, Y. and Zhang, L. (2021). Roles of artificial intelligence in construction engineering and management: a critical review and future trends. *Automation in Construction,* 122(August 2020), 103517. https://doi.org/10.1016/j.autcon.2020.103517.

Štefanič, M. and Stankovski, V. (2019). A review of technologies and applications for smart construction. *Proceedings of the Institution of Civil Engineers: Civil Engineering,* 172(2), 83–87. https://doi.org/10.1680/jcien.17.00050.

Suleiman, A. and Ayer, S.K. (2019). Mobile augmented reality to influence design and constructability review sessions. *Journal of Architectural Engineering,* 25(3), 04019016. https://doi.org/10.1061/(ASCE)AE.1943-5568.0000362.

Tao, F., Cheng, J., Qi, Q., Zhang, M., Zhang, H. and Sui, F. (2018). Digital twin-driven product design, manufacturing and service with big data. *International Journal of Advanced Manufacturing Technology,* 94(9–12), 3563–3576. https://doi.org/10.1007/s00170-017-0233-1.

Wu, C., Li, X., Guo, Y., Wang, J., Ren, Z., Wang, M. and Yang, Z. (2022). Natural language processing for smart construction: current status and future directions. *Automation in Construction,* 134(October 2021), 104059. https://doi.org/10.1016/j.autcon.2021.104059.

Yan, H., Yang, N., Peng, Y. and Ren, Y. (2020). Data mining in the construction industry: present status, opportunities, and future trends. *Automation in Construction,* 119(August 2019), 103331. https://doi.org/10.1016/j.autcon.2020.103331.

Zhang, F., Chan, A.P.C., Darko, A., Chen, Z., & Li, D. (2022). Integrated applications of building information modeling and artificial intelligence techniques in the AEC/FM industry. *Automation in Construction*, 139, 104289. https://doi.org/10.1016/j.autcon.2022.104289

3

Digital Twins and Sustainable Development Goals

Noha Saleeb

Middlesex University, London, UK

Introduction

The concept of digital twins in construction has been controversial in its definition, since some consider it to be a virtual representation or model of a physical asset (Kaur et al. 2020), while others extend it to a bi-directional flow of information between the physical and digital. For example, through collection of different types of sensors, telemetry data can be captured. These data are then quantitatively and qualitatively analysed in the virtual environment and fed back into the physical environment, as identified by Madubuike et al. (2022) and Lucchi (2023). Furthermore, there has been considerable debate regarding the levels of digital twin functionality to aspire to implement for different assets. For example, Lueth (2020) defines 250 classifications for digital twins based on three hierarchical axes: (i) Six hierarchical levels of objects (informational, component, product, process, system and multi-system), (ii) six levels of usage (digitise, visualise, simulate, emulate, extract, orchestrate and predict) and (iii) six life cycle stages (design, build, operate, maintain, optimise and decommission). A Sustainability Digital Twin Framework may be created that is suitable for any of these levels. Another view by Kharche (2022) defines a further five maturity levels for a digital twin: (i) Foundation (real-time data collection), (ii) predictive (performance management and analytics), (iii) prescriptive (optimisation and social), (iv) transformative (collaboration and training) and (v) cognitive (autonomous operations, Artificial Intelligence [AI], robotics, etc.). However, the perspective of this study envisions the feasibility of incorporating certain aspects of all five of these maturity levels simultaneously, for example, the analysis of live data feeds for optimisation of operations to achieve indoor climate comfort with AI statistical predictability. Current digital twin projects are primarily focused on enhancing efficiency, reducing waste and improving project management. Yet, aligning these advancements with the United Nations' Sustainable Development Goals (UN SDGs) involves recognising how these technologies can contribute to broader sustainability objectives. In 2015, 17 UN SDGs were adopted globally instead of the previous 8 Millennium Development Goals, in an attempt to include a wider scope of sustainability

Architectural Design and Management in the Digital Age: International Perspectives, First Edition.
Edited by Bob Giddings, Silvio Melhado, and Jenni Barrett.

within all disciplines, as reported by the United Nations (2023a). Since then, there have been numerous endeavours to align digital twin projects with the UN SDGs. Nevertheless, the majority of these projects have connected construction directly to only seven SDGs (Tzachor et al. 2022). These are SDG 3 Good Health and Well-being, which is directly associated with indoor climate design; SDG 6 Clean Water and Sanitation, which is directly related to enhancing water services and systems; SDGs 7, 11 and 12 Affordable and Clean Energy, Sustainable Cities and Communities and Responsible Consumption and Production, which are connected to reducing energy consumption, carbon footprint, waste and enhancing circularity; SDG 9 Industry Innovation and infrastructure, which relates to new construction technologies for better performance and SDG 13 Climate Action, which by association would be aligned to a broad range of attributes. There are insufficient research and development attempts to align the other 10 SDGs as a priority outcome of digital twin projects in the construction industry, and even at a city level, the scope is limited to just a few SDGs (Koeva et al. 2023). These tend to include SDGs 1, 2, 4 and 5 No poverty, Zero Hunger, Quality Education and Gender Equality, which are considered by many practitioners to be only social goals; SDGs 8 and 10 Decent work and economic Growth and Reduced Inequalities, which are perceived by many observers as primarily national macro-level economic goals and not as micro-level organisations or project goals; SDGs 14 and 15 Life below Water and Life on Land, which are considered as mainly natural environmental goals and not built environment related and SDGs 16 and 17 Peace, Justice and Strong institutions and Partnerships for the Goals, which are seen as political aims. The challenge that needs to be addressed is how to incorporate all 17 SDGs within the design and implementation of a digital twin, and the key performance indicators (KPIs) involved.

Current Digital Twins Aligning with UN SDGs

As previously mentioned, the alignment of digital twins with SDGs often revolves around leveraging technology to promote sustainable practices, resource optimisation, waste reduction and improved decision-making processes in specific 6–7 SDGs. Examples where digital twins contribute to particular SDGs are as follows:

1. Sustainable Building Design and Operation: Digital twins can facilitate the design, construction and operation of sustainable buildings. Integration of data analytics, simulations and real-time monitoring can optimise energy efficiency, reduce waste and improve building performance to align with SDG 7 – Affordable and Clean Energy, SDG 9 – Industry, Innovation and Infrastructure and SDG 11 – Sustainable Cities and Communities. An example is the usage of digital twin within Amsterdam Airport Schiphol, a 7000 acre complex, to optimise operations and for predictive maintenance, through analysis of model data, geographic information system (GIS) data, real-time remote sensor changes and financial records. This allows monitoring and running 80,000 assets from linear networks, runways, and lighting systems to information booths and fire extinguishers by conducting simulations on potential operational failures to reduce costs and time (Baumann 2019).

2. Resource Efficiency and Waste Reduction: Utilising digital twins to manage resources effectively, minimise material wastage and streamline construction processes, aligns with SDG 12 – Responsible Consumption and Production. An example is Florida Utility of its Electric Assets for highly efficient field operations and resourcing. The digital twin coupled with live telemetry from 2 Arrow Gold receivers allowed accurately depicting locations of all operations and transmissions and distribution assets. Thus, all field operations are faster, more efficient and better resourced with less waste in electric and water services (ESRI 2023).
3. Resilient Infrastructure Development: Digital twins can aid in creating and maintaining resilient infrastructure, which is crucial for achieving SDG 9 – Industry, Innovation and Infrastructure and SDG 13 – Climate Action. By modelling and simulating various scenarios, these twins can help in designing infrastructure that can withstand environmental challenges and natural disasters. An example is the Caribbean Island of Grenada where predicting and simulating natural disasters enable mitigation plans to help agriculture and tourism, which are the main sources of income for the island. The digital twin uses live telemetry in Light Detection and Ranging remote sensing imagery data (LiDAR) imagery data to produce layers of island mapping and analysis of natural and artificial built resources, e.g. vegetation, water courses, roads, buildings and power lines (Chadha 2022).
4. Data-Driven Decision-Making: Leveraging the power of data allows for informed decision-making, fostering efficiency and sustainability across the construction life cycle. This aligns with SDG 9 – Industry, Innovation and Infrastructure and SDG 17 – Partnerships for the Goals. An example is the partnership between Gwinnett County Department of Water Resources USA and KCI to develop a digital twin to enforce better decision-making over 200 pump stations and 8000 miles of sewer, clean and storm water piping systems. This digital twin incorporated LiDAR and sensor telemetry data implemented over its whole infrastructure. This allows reduction of maintenance issues, extended life cycles, safety optimisation, accurate location and specifications (Informed Infrastructure 2023).
5. Reduction of Carbon Emissions: Implementing digital twins in construction allows for better planning, which can minimise transportation needs, optimise energy consumption and reduce the carbon footprint of construction projects. This aligns with SDG 13 – Climate Action and SDG 11 – Sustainable Cities and Communities. An example is the digital twin created by the city of Uppsala in Sweden primarily for achieving sustainable goals and reduction of carbon emissions. The city's green approach became world leading with a pledge to be fossil-fuel free by 2030 and climate positive by 2050. The Uppsala Climate Protocol and Roadmap collaborates with 42 local businesses, public agencies, associations and the city's two universities to meet its climate challenge, and create a guide on how to phase out fossil fuels. The digital twin aims to provide solutions for sustainable living by the citizens to manage their built spaces and simulate usage of different renewable materials and resources (Patrick 2021).
6. Enhancing Safety and Thermal Comfort: Digital twins enable better monitoring and simulation of safety protocols, predicting and reducing accidents by analysing precursor actions or activities on construction sites, enhancing inspection and visualisation accuracy and improving worker safety. They can also help to improve conditions for users while living or working in an indoor climate (Park et al. 2023a). This aligns with

SDG 3 – Good Health and Well-being. An example is the digital twin created for a Physiological Equivalent Temperature Calculation to accurately assess thermal comfort in spaces as a metric for heat stress. This is by identifying the effect of adding or removing trees, buildings, water and green areas. There is the simulation of micro-climates, including urban air temperatures and wind speeds using routine weather observations and information about land use and building geometries. Its goal is to achieve target temperatures and wind speeds in key urban areas, and for decision-makers to understand solutions, as used in the Dutch city of Enschede (Cárdenas et al. 2023; Koeva et al. 2023) and others.

7. Community Engagement: By utilising digital twins, stakeholders can visualise and understand the impact of construction projects on local communities, encouraging transparent and participatory planning. This aligns with SDG 11 – Sustainable Cities and Communities. Abdeen et al. (2023) state that data collection for digital twins should comprise Human Sensors Data, which include emergency situational user data, real-time feedback and perception of citizens and stakeholders, users' geographical information, behavioural data and preferences.
8. Long-Term Infrastructure Planning: Digital twins facilitate better long-term planning, maintenance and management of infrastructure, contributing to sustainable and resilient development. This aligns with SDG 9 – Industry, Innovation and Infrastructure and SDG 17 – Partnerships for the Goals. An extensive case study is the South Korean comprehensive endeavour to implement digital twins within several of its major cities for long-term planning. For example, Incheon, a major industrial city, implemented a digital twin project that includes monitoring and optimising multiple major areas – fire response management, traffic, urban sanitation, facilities management, urban development, city revitalization, waste disposal, disease control in connection with buildings, power grids and telecommunications management, flood prediction monitoring and users' experiences. Internet of Things (IoT) sensors, GIS, mobile network databases, municipalities' data and surveys, 3D point clouds, public vehicles' GPS tracking systems are all used in tandem for data collection and analysis. It is estimated that the digital twin has increased efficiency among Incheon city government offices by 10% (Milner 2021). Another example is the digital twin created for Jeonju city for land management, purposing, maintenance, planning and human resource management for different services. This involved collaboration between the Ministry of Land, Infrastructure and Transport and other public and private entities for data collection, GIS, surveys, maps, 3D creation, database formation and data analysis (Park et al. 2023b).

Other digital twin technical, social and economic advantages include – locating building, level, room, components inventory and occupancy; implementing scheduled and preventative Maintenance Plan; compliance assistance for State and Federal Funding; commissioning and decommissioning; insurance benefits; remote workspace and learning; facilities Condition Assessments and Risk Management Reviews; Real Estate and Capital Planning and Simulation Prototypes for future spaces and asset needs. However, it can be seen from all the examples that the UN SDGs addressed by existing construction digital twins mainly focus on SDGs 3, 7, 9, 11, 12, 13 and 17, with no indicated alignment with the other SDGs. Hassani et al. (2022) give examples of further SDGs that could be relevant to different areas of digital twin applications. For example, sustainable production, smart manufacturing, energy, resourcing and construction, product life cycle management, building maintenance, logistics and circular economy could align to SDG 1 – No Poverty, SDG 8 – Decent Work and Economic Growth, and SDG 10 – Reduced Inequalities, through providing more specialised employment opportunities for all genders of the population. This is in addition to the expected SDGs 6, 7, 9, 11 and 12 related to Clean Water and Energy, Industry Innovation, Sustainable Cities and Responsible Consumption. Another example is the possible alignment of smart agriculture, farming, fishery, livestock to SDG 2 – Zero Hunger, SDG 14 – Life below Water and SDG – Life on Land. Yet, the research does not explain how these can connect to the built environment, but it can be inferred that designing suitable farms and factories that would not eliminate agricultural land or pollute waters or eliminate forests and animal natural habitats could be a consideration. Furthermore, Giret et al. (2023) recommend the use of AI to supported digital twin applications, in three layers as follows: (i) the Data layer combines data from different sources, (ii) the Model layer acts as a physically consistent integrator of the information from the previous layer in four dimensions of space and time, (iii) the Decision Support layer can deduce future actions and events, by merging outputs from the Model layer with simulation engines, expert knowledge and integration with external systems. AI would provide advanced pre-processing and post-processing analytics, hybrid simulation, augmentation, data fusion, mining and discovery, recommendations, reasoning and scenario generation. An example is the use of digital twins in Earth Observation Systems for sustainability. In Europe, this is championed by the European Commission's Green Deal Strategy. This is through the Destination Earth initiative to develop high-precision digital models of the Earth, or digital twin of the Earth, offering a digital modelling platform to visualise, simulate, monitor and forecast impacts of natural and human activity and guide informed decisions for sustainable development.

It should be noted that some of the national initiatives discussing digital twins adopted by the construction industry globally do not delve into the alignment with the SDGs. The UK BIM Alliance (2022) positioning statement report on BIM and digital twins only discusses the perception of digital twins based on the ISO19650 suite of standards with no indication to sustainability integration. The Centre for Digital Built Britain (2022) also produced a series of reports as part of the Digital Twin Hub Gemini Project reports. This Project advocates the concept of connected digital twins, alluding to high-level economic, social, digital and natural environment infrastructures, but with no details of sustainable categories, which are also not included in the Gemini principles.

SDGs Key Performance Indicators in Construction Digital Twins

In order to align construction industry digital twins to sustainable practices, it is essential to first identify existing sustainability KPIs or standards available globally, to investigate the possibility of utilising them or customising them for usage in the built environment. A most recent Report of Global SDGs Indicators produced by the United Nations (2023b) defines in detail tier classifications for the KPIs identified by the UN for countries and organisations to report on their initiatives and degree of achieving each of the 17 SDGs. This report divides all KPIs into three tiers. As mentioned in the reply box, it would be good not to use acronyms all the timeTier 1 refers to KPIs that have well-identified metrics and are currently being reported with data being provided for them from at least 50% of countries. Tier 2 refers to clearly identified KPIs in terms of metrics, standards and methodologies, yet there is insufficient data being provided for them from countries around the world. Tier 3 refers to KPIs that have unclear metrics, methodologies and standards. As can be seen in Figure 3.1, the report is in tabular form identifying each SDG goal, its categories of assessment, the KPIs for each category, their tier classification and custodians and partner organisations. Figure 3.1 identifies some of the major KPIs in each SDG that can be customised by digital twins in the construction industry to align to each SDG.

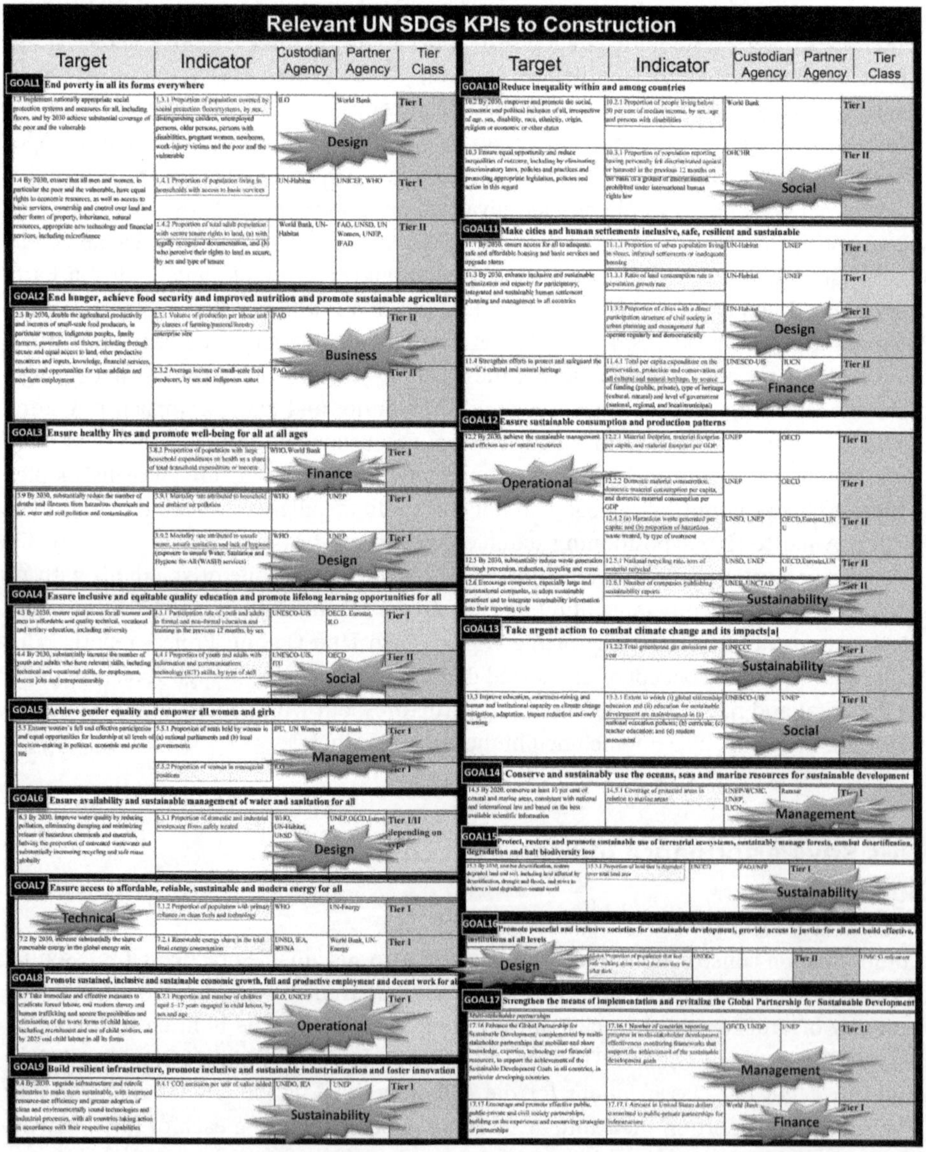

Figure 3.1 Alignment of UN SDG key performance indicators with digital twin disciplines.

For SDG 1, KPI 1.3.1 assesses the proportion of the population who have housing as social protection. Digital twins can identify through occupancy sensors and digital records, the percentages of different categories of users occupying different spaces, floors, buildings and compare these statistically to assess and recommend levels of poverty reduction. This includes analysing the proportion of these housing facilities with basic services, their change over time (KPI 1.4.1) and how many of these assets are tenured, loaned, rented, etc. (KPI 1.4.2). These KPIs require design efforts to create more houses for different economic standards of residents. In SDG 2, digital twins can be used to measure production levels of factories (KPI 2.3.1) to aid in distribution of resources where needed. For SDG 4, digital twins can evaluate the percentage of different ages at different stages of education, and also specifically the proportions in different specialised disciplinary areas (KPI 4.3.1 and 4.4.1). As well as quantitative statistics, these KPIs require social qualitative data analyses to assess quality of education. It is also essential to create more training courses on sustainable use of buildings that can be disseminated to industry practitioners and users and can be monitored using digital twins. In SDG 5, KPIs work closely in tandem with SDG 4, where a high-level managerial strategy needs to be formulated to allow equal gender representation in different industry roles, the percentages of which can be assessed using digital twins to analyse public and company databases (KPI 5.5.1 and 5.5.2). The KPIs in SDG10 adopt a social context for identifying percentages of people below 50% median salary range. These statistics can be shown using dashboards for GIS locations using digital twins (KPI 10.2.1). This can help identify incidents of discrimination, within the workplace (KPI 10.3.1).

The Technical SDGs, i.e. SDG 3 – Good Health and Well-being, SDG 6 – Clean Water and Sanitation, SDG 7 – Affordable and Clean Energy, SDG 9 – Industry Innovation and Infrastructure, SDG 11 – Sustainable Cities and Communities and SDG 12 – Responsible Consumption and Production. Figure 3.1 shows that for SDG 3, digital twins can calculate different proportions of household expenditures on services, such as heating, cooling, lighting, humidity, insulation and to achieve indoor comfort, in comparison to overall household expenditure (KPI 3.8.2). Rates of indoor air pollution can be assessed using air pollution sensors (KPI 3.9.1) and also energy consumption, carbon emissions, water pollution using sensors, and hence the correlation with users' health and well-being is identified (KPI 3.9.2). In SDG 6, it is essential to design clean and wastewater systems in buildings to comply with global health standards. These can be assessed using water quality sensors and analysed within a digital twin (KPI 6.3.1). SDG 7 and SDG 9 require sustainable technical design solutions to increase adoption of clean and renewable energies by users to reduce reliance on fossil fuels for generating electricity and also reduction of carbon emissions. Measuring the proportion of renewable and non-renewable energy, and also resultant carbon emissions, can be calculated with digital twins using live telemetry data collected from carbon and energy sensors to achieve KPIs 7.1.2, 7.2.1 and 9.4.1. SDG 11 delves into the macro level of assessing whole communities' and cities' resilience. This requires digital twins to record building data, and identify percentages of slums, informal settlements and other low standards, in order to formulate, design and finance holistic solutions for their elimination or upgrading (KPI 11.1.1). Furthermore, by identifying ratios and rates of land consumption to dynamic population growth in different locations using digital twins, this can help future urban planning (KPIs 11.3.1 and 11.3.2). Also identifying buildings and spaces for conservation and heritage using digital twins will help assess the financial resources for their preservation (KPI 11.4.1). Finally, in SDG 12, KPIs can be achieved through statistical analysis performed with the aid of digital twins and live telemetry data to calculate embodied and operational carbon footprint, and also the amount of waste generated in different periods and locations, which can help formulate a logistic strategy for material resourcing and also waste reuse, recycling and disposal.

Environmental SDGs, i.e. SDG 13 – Climate Change, SDG 14 – Life Below Water and SDG 15 – Life on Land. For SDG 13, the crucial KPI, to align to climate change assessment, is the calculation of total greenhouse emissions per year. This can be calculated by merging carbon calculators with digital twins to capture all changing live attributes, conditions and data across time (KPI 13.2.2). As previously mentioned, attention to training and educating people about sustainable solutions is crucial (KPI 13.3.1). While SDG 14 and SDG 15 initially appear to be the most inapplicable for digital twins and the construction industry, in reality, they are crucial in enabling buildings to be designed to eliminate pollutants and waste (KPI 14.5.1). The same applies to conserving land and forests, and not just transforming them into plots for building (KPI 15.3.1). Dynamic water contamination information can be collected using live sensor telemetry data and assessed using digital twins. Furthermore, digital twins can be used to calculate the percentage of land utilised for different purposes and how the use overlaps with wildlife habitats. The Economic SDGs are SDG 8 – Decent Work and Economic Growth, SDG 16 – Peace Justice and Strong Institutions and SDG 17 – Partnerships for the Goals. In SDG 8, a critical KPI is to assess legal and illegal employment. The latter can include underage workers and modern slavery. In these circumstances, it is difficult to achieve accurate metrics but digital twin databases can be updated with indications from charities and aid organisations (KPI 8.7.1). SDG 16 advocates peaceful living by the population and feeling safe at night while walking in the neighbourhood (KPI 16.1.4). This requires survey data with citizens identifying safety in different places. Once outlined, this can be plotted using a digital twin to help design safer environments. Finally, SDG 17 encourages the formation of partnerships to help achieve sustainability goals. In terms of digital twins, this requires financed collaboration and management between government, technology providers, operational institutions and users (KPI 17.16.1 and 17.17.1). Based on all previous information pertinent to the importance of digital twins and identifying suitable SDG KPIs for implementation, a framework will be identified for creating a design and construction digital twin that embeds all the SDGs and aligns with its KPIs.

Framework for Embedding UN SDGs Within a Digital Twin

Creating a design and construction digital twin that aligns with the SDGs' KPIs requires planning in four stages. A framework can be proposed for implementing a consolidated digital twin (Castaldini 2019; Deloitte 2017; Madubuike et al. 2022; Tao et al. 2018). Additionally, a live case study is presented as an example of implementing the framework to create a digital twin aligning with all SDGs (Saleeb 2023). The main aim is to create a renovation system and classification of a heritage-listed castle in Egypt as a model to apply to other historical assets in different countries, to improve these invaluable structures, in a bid to improve tourism to these places, increase employability and ultimately increase a country's Gross Domestic Product (GDP).

Stage A – Data Acquisition. This stage involves preparing documentation, databases and all information available about a design. There are two steps involved in this stage:

1. Defining data structures and systems using extensive project documentation, for example, BIM execution plans, task information delivery plans, mobilisation plans, BIM models, as-built models, asset information requirements, maintenance manuals, spreadsheets, existing databases in different systems, council or registry information.
2. Defining levels of detail, levels of development and levels of information required, including purpose of the information, format, function, etc. This is crucial to create a digital twin virtual representation to the exact details required for data analysis.

Stage B – Data Preparation and Analysis. This stage involves processing all acquired data in the previous stage to be ready for inclusion within the digital twin. This includes cleansing, formatting, verifying the data against standards, validating the data for usage and ensuring interoperability of all types of data. There are two steps involved in this stage:

3. Creating a library database to connect different nongraphical data with the digital twin graphical data.
4. Defining industry standards and regulations with which to adhere and comply. These can be national or global standards and are especially important for international projects.

Stage C – Digital Twin Design. This stage involves creation of the digital twin structure and defining all its requirements from human resources to software and hardware resources. There are three steps involved in this stage:

5. Defining roles and skill sets required to work on the digital twin project. An example from the FRESH-DT project (Saleeb 2023) revealed the need for a whole host of diversified skill sets from architectural designers to computer scientists, to social and environmental scientists. The roles identified were BIM specialist, software programmer, asset manager, data analyst, visual analyst, structural engineer, architect, interior designer, mechanical, electrical and plumbing (MEP) engineer, researchers, renovation specialist and environmental specialist.
6. Aligning the digital twin objectives with each SDG, its objectives and KPIs for assessment within the Digital Twin. This step is crucial to describe how the different SDG KPIs will be assessed and achieved in tandem with the digital twin project KPIs. Within the FRESH-DT project (see Figure 3.2), each of its five project outputs were aligned with certain SDGs and their KPIs. For example, the newly created Heritage Classification System (Saleeb 2023) was identified as achieving industry innovation (SDG 9) to promote a more sustainable community (SDG 11), while in order to implement it, partnerships with government and public professional bodies are significant as enablers (SDG 17). The Framework for Structural Analysis would help achieve health and safety for users of the building (SDG 3) in addition to providing an innovative industry solution (SDG 9). The Expert Renovation System helped improve the quality of water services (SDG 6), helped reduce energy consumption and carbon emissions (SDG 11). Also, it would allow more job specialisms and opportunities for professionals in design and construction (SDG 8) and would be suitable for training for other types of heritage assets (SDG 4). The Virtual Reality and Augmented Reality simulations of social behaviour of users inside the building would help

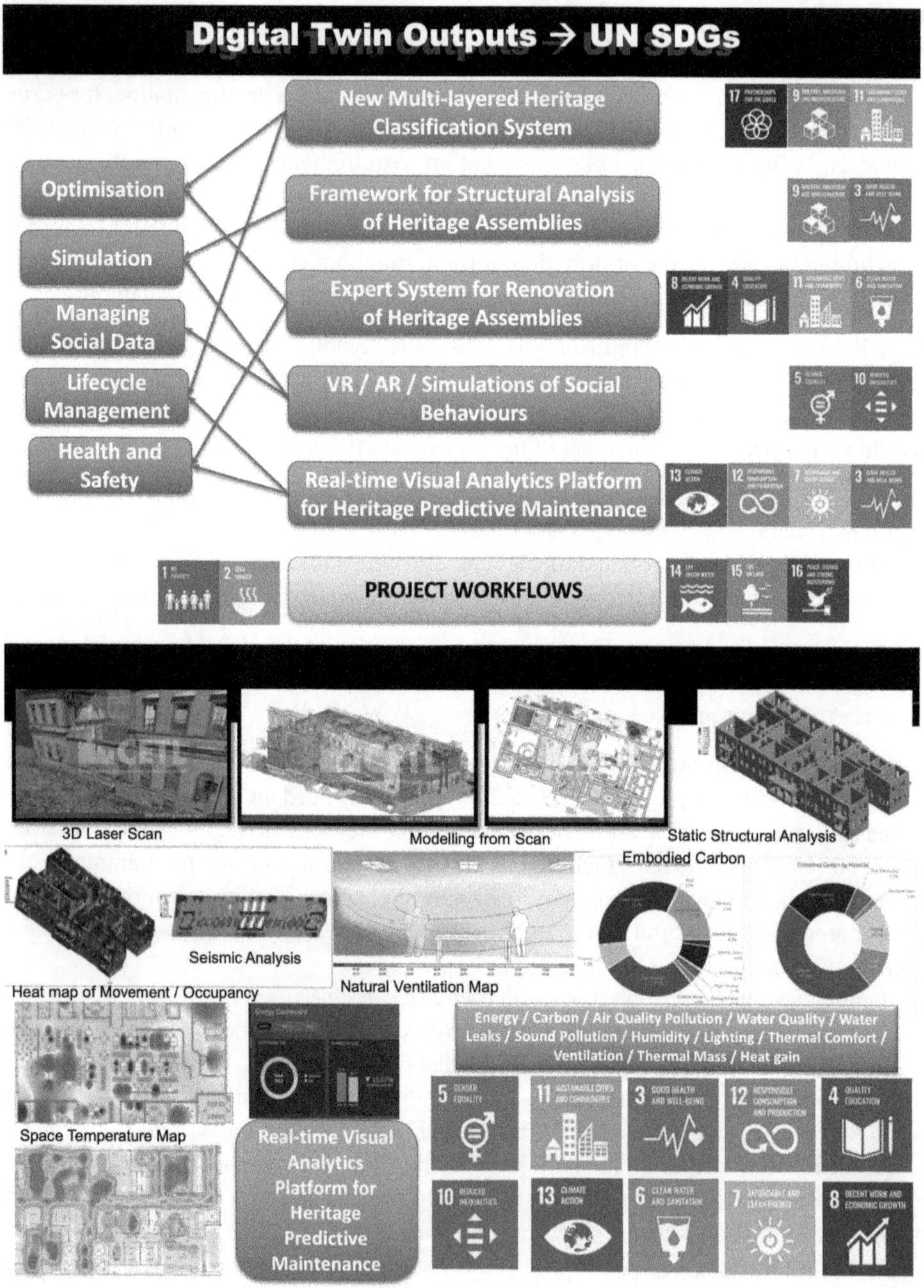

Figure 3.2 Digital twin outputs and UN SDG goals.

decision-makers design activities and crowd flow based on inclusivity and diversity of genders and people with disabilities (SDG 5 and 10). The Real-Time Visual Analytics Platform showcasing the building performance in many operational areas would allow better decision-making for maintenance and operations ensuring better users' health and well-being in the indoor climate (SDG 3). It would also provide more information to compare materials, techniques for renovation, energy sources that are renewable and sustainable regarding both embodied and operational carbon to help fight climate change (SDG 7, 12 and 13). Finally, some operations while conducting the FRESH-DT helped to achieve the following SDGs. This project opens more opportunities for employment in the tourism sector thus helping to fight poverty (SDG 1). Furthermore, special attention was allocated to improving the drainage system to eliminate harmful waste and pollution, since the system directly connected to the natural water way and would harm any underwater organisms (SDG 14). In addition, all bats and snakes that were infesting the building were safely removed and relocated to a different habitat (SDG 15). Finally, upon completion of this project, several partnerships were formed with several professional bodies and as a result, the output systems, e.g. the Heritage Classification System, can be utilised on other major heritage projects (SDG 16, 17).

7. Creation of digital twin architecture, including any required software platforms, hardware equipment for high processing, analysis and archiving capabilities and any other necessary technology platforms for analysis, e.g. usage of machine-learning algorithms or big data analysis. Figure 3.3 outlines the system architecture used for the FRESH-DT project. Sixteen different types of sensors were used to collect live telemetry data from within different sized and functional spaces in the heritage asset investigated. These sensors were then transferred the data to the digital twin to enable analysis.

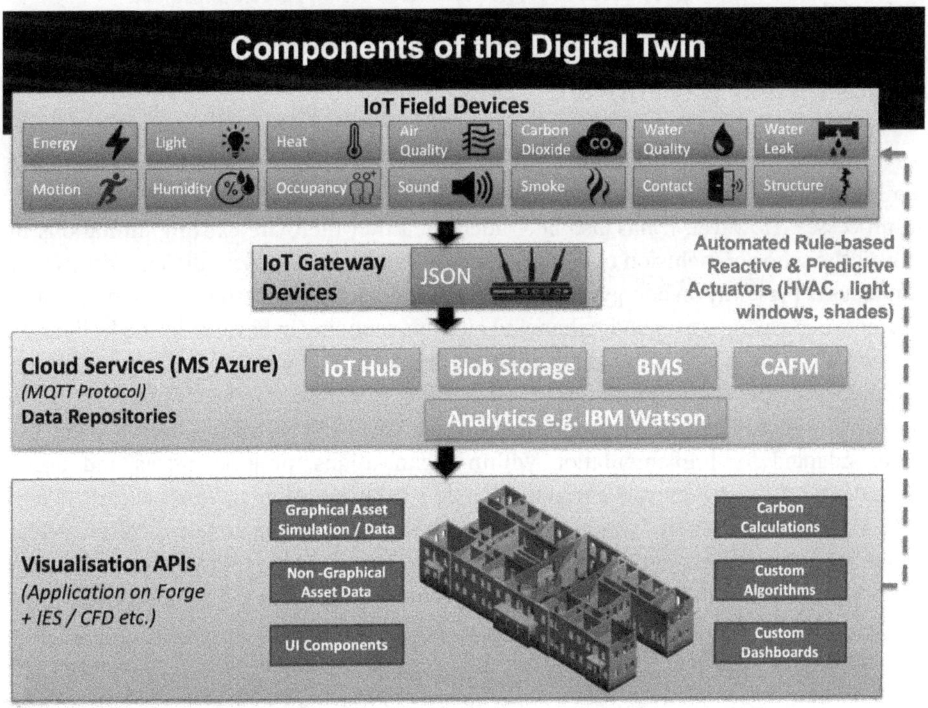

Figure 3.3 Components of digital twin system architecture.

Stage D – Digital Twin Implementation. This stage involves setting up the digital twin by creating the virtual models and connectivity, and also implementing the sensor devices and connecting equipment on site trialling the system and running it. There are three steps involved in this stage:

8. The first step involves choosing suites of compatible sensors to implement within the asset, deciding on the manner of installation, including wall, floor, ceiling, etc. Also, whether the sensors are charged by electricity or battery or both for emergencies needs to be assessed. Whether the sensors transmit their data via cables or remotely via Wi-Fi is another issue. Any racks, cabling, appliances required for installation and maintenance are evaluated. The visual impact of these sensors on the internal spaces is a significant matter. The frequency of maintenance should be scheduled. The availability of online servers available for data collection or the need to book cloud space should be identified.
9. Creating the Digital Twin and Dashboards. This step is implementing what was designed in the architecture of the system. It is important to define a timeline for installing all the different components. One of the challenges faced in the FRESH-DT project was obtaining governmental permissions for installing, collecting and archiving data about a public asset.
10. Piloting, Scaling and Data collection. This is the implementation final step which starts with conducting a pilot to test the system performance, which would then be adjusted to account for any errors before scaling the pilot to the full asset and proceeding with the data collection, analysis and display of results. The dashboards should also include the assessed metrics of the SDG KPIs achieved by the project.

Conclusion and Way Forward

Current industry practices in terms of existing digital twin projects and endeavours have been highlighted, with focus on their current functional goals and enhanced workflows and processes. However, it has also been identified that there are existing limitations in terms of the scope of inclusion of sustainable goals addressed by these digital twins. Their main focus is presently on energy consumption reduction, lowering the carbon footprint of buildings, decreasing waste and enhancing circular economy behaviour in the built environment, but there is lack of insight to align too many of the UN SDGs. Furthermore, there are scarce attempts to customise the UN SDG KPIs, as defined by the United Nations, to adopt within design and construction. Examples of the SDG KPIs have been shown that can be adapted for implementation within organisations, projects, assets and users. Subsequently, a framework was introduced on how to create a digital twin that can align to all the SDGs, with a case study example of live site implementation and work performed to include KPIs as an assessment mechanism for the value of the digital twin case study. The way forward and future research should endeavour to implement more industry 4.0 technologies within design and construction digital twins as a catalyst to enhance sustainable productivity, for example, (i) the inclusion of cognitive computing which includes, AI Machine Learning coupled with Natural Language Processing, which can aid in the prediction of social behavioural patterns that can influence sustainable outcomes, and hence provide informed suggestions on how to enhance them or expect risky actions and mitigate them. (ii) The adoption of Block Chain Technologies can help in securely archiving, transmitting and exchanging information in an immutable and fast way that would encourage more users to share information through a digital twin and also provide more trust in the results and analyses of the digital twin dashboards, decisions and suggestions. In addition, smart contracts could be utilised to calculate carbon emissions and carbon credits as incentives for reduction of carbon footprint by both organisations and individuals constructing or utilising buildings. (iii) The adoption of autonomous drones and robotics can assist in monitoring inaccessible or dangerous areas, regions for humans enforcing actions that can support building sustainable functions and health and safety. (iv) The implementation of Big Data Analytics can assist in structuring the enormous amount of live data collected from live telemetry sensors and aid in the mining of relevant information to enhance performance and sustainability of buildings. This would be coupled with visual analytics dashboards that show live data streams and their immediate impact on different sustainability measures. (v) Virtual and Augmented Reality applications can help in training and raising awareness among different user genres on how to raise their sustainability practices within their buildings. (vi) Cloud Computing is a crucial resource to allow all previous technologies to have enough processing capability and power to perform seamlessly, quickly and in an efficient interoperable fashion. As noted by Oláh et al. (2020), these technologies are currently primarily in alignment with UN SDGs 7, 9, 12 and 13, and hence it is important to broaden their implementation to include other SDGs, since they do not only support sustainable outcomes for environmental goals but also technical, social and economic goals. This would include smart grids, life cycle assessments, enhanced prototyping, reduced costs, easier compliance with standards, improved employability and productivity and promoting intelligent learning analysis. It is therefore significant that decision-makers and policy-makers within the construction industry revisit their view of sustainability to a wider lens that encompasses the integration of the built environment in all areas of future sustainability.

References

Abdeen, F.N., Shirowzhan, S. and Sepasgozar, S.M., (2023). Citizen-centric digital twin development with machine learning and interfaces for maintaining urban infrastructure. *Telematics and Informatics*, 84, 102032.

Baumann, J., (2019). Digital twin helps airport optimize operations. *Esri*. https://www.esri.com/about/newsroom/wp-content/uploads/2019/10/airport.pdf (accessed 10 December 2023).

Cárdenas, I.L., Morales, L.R., Koeva, M., Atun, F. and Pfeffer, K. (2023). Digital twins for physiological equivalent temperature calculation guide. Zenodo. https://doi.org/10.5281/zenodo.8306385.

Castaldini, F. (2019). How digital twin was central to smart buildings. *Facility Executive*. https://facilityexecutive.com/2019/02/how-digital-twin-technology-is-central-to-smart-buildings/ (accessed 10 December 2023).

Centre for Digital Built Britain. (2022). Gemini papers: what are connected digital twins? https://www.cdbb.cam.ac.uk/files/gemini_papers_-_what_are_connected_digital_twins.pdf (accessed 10 December 2023).

Chadha, P. (2022). Climate change prompts Caribbean island to create its digital twin. https://www.geospatialworld.net/prime/technology-and-innovation/climate-change-prompts-caribbean-island-to-create-digital-twin-of-entire-country/#:~:text=Grenada%20used%20its%20countrywide%20LiDAR,imagery%20and%20point%2Dcloud%20data (accessed 10 December 2023).

Deloitte. (2017). Industry 4.0 and the digital twin – Manufacturing meets its match. *A Deloitte series on Industry 4.0*, Deloitte University Press.

ESRI. (2023). Case study: Florida utility creates digital twin of electric assets with highly accurate field operations. https://www.esri.com/en-us/landing-page/industry/electric-and-gas/2020/new-smyrna-beach-case-study (accessed 10 December 2023).

Giret, A., Julian, V. and Carrascosa, C. (2023). AI-supported digital twins in applications related to sustainable development goals. *Proceedings of The International FLAIRS Conference, (May 14–17, 2023), Clearwater Beach, Florida*. Vol. 36.

Hassani, H., Huang, X. and MacFeely, S. (2022). Enabling digital twins to support the UN SDGs. *Big Data and Cognitive Computing*, 6(4), 115–131.

Informed Infrastructure. (2023). Gwinnett county explores benefits of digital twins at pump station. https://informedinfrastructure.com/gis-solutions-gwinnett-county-explores-benefits-of-digital-twins-at-pump-station/ (accessed 10 December 2023).

Kaur, M.J., Mishra, V.P. and Maheshwari, P. (2020). *The convergence of digital twin, IoT, and machine learning: transforming data into action. Digital twin technologies and smart cities.* New York: Springer. 3–17.

Kharche, V. (2022). Digital twin maturity. https://www.linkedin.com/pulse/digital-twin-maturity-vinayak-kharche/ (accessed 10 December 2023).

Koeva, M., Cardenas, I.L., Morales, L.R., (2023). Digital twins for a sustainable world. Developments, challenges and applications of city digital twins in the context of the SDGs. https://www.gim-international.com/content/article/digital-twins-for-a-sustainable-world (accessed 10 December 2023).

Lucchi, E. (2023). Digital twins for the automation of the heritage construction sector. *Automation in Construction*, 156, 105073. https://doi.org/10.1016/j.autcon.2023.105073.

Lueth, K.L. (2020). How the world's 250 digital twins compare? Same, same but different. *IoT Analytics*. https://iot-analytics.com/how-the-worlds-250-digital-twins-compare/ (accessed 10 December 2023).

Madubuike, O.C., Anumba, C.J. and Khallaf, R. (2022). A review of digital twin applications in construction. *Journal of Information Technology in Construction*, 27, 145–172. https://doi.org/10.36680/j.itcon.2022.008.

Milner, G. (2021). South Korean city uses a digital twin to meet challenges. https://www.esri.com/about/newsroom/arcuser/south-korean-city-uses-a-digital-twin-to-meet-challenges/ (accessed 10 December 2023).

Oláh, J., Aburumman, N., Popp, J., Khan, M.A., Haddad, H. and Kitukutha, N. (2020). Impact of industry 4.0 on environmental sustainability. *Sustainability,* 12(11), 4674.

Park, J.S., Lee, D.G., Jimenez, J.A., Lee, S.J. and Kim, J.W. (2023a). Human-focused digital twin applications for occupational safety and health in workplaces: a brief survey and research directions. *Applied Sciences*, 13(7), 4598.

Park, J., Choi, W., Jeong, T. and Seo, J., (2023b). Digital twins and land management in South Korea. *Land Use Policy*, 124, 106442.

Patrick, B. (2021). Uppsala creates a detailed digital twin to enhance sustainability. https://www.esri.com/about/newsroom/blog/uppsala-sustainable-development/ (accessed 10 December 2023).

Saleeb, N., (2023). Framework for restoration of heritage using digital twin (FRESH-DT). *London Digital Twin Research Centre, Future of Digital Technologies.* https://dt.mdx.ac.uk/?page_id=1268 (accessed 10 December 2023).

Tao, F., Sui, F., Liu, A., Qi, Q., Zhang, M., Song, B., Guo, Z., Lu, S.C.Y. and Nee, A.Y.C. (2018). Digital twindriven product design framework. *International Journal of Production Research,* 57(12), 3935–3953.

Tzachor, A., Sabri, S., Richards, C.E., Rajabifard, A. and Acuto, M., (2022). Potential and limitations of digital twins to achieve the sustainable development goals. *Nature Sustainability*, 5(10), 822–829.

UK BIM Alliance. (2022). BIM and digital twins UK BIM alliance positioning statement report. https://ukbimalliance.co.uk/wp-content/uploads/2021/06/UKBIMA_BIM_DigitalTwins.pdf (accessed 10 December 2023).

United Nations. (2023a). The Sustainable Development Goals Report 2023: Special Edition. https://unstats.un.org/sdgs/report/2023/ (accessed 10 December 2023).

United Nations. (2023b). Tier classification for global SDG indicators report. https://unstats.un.org/sdgs/files/Tier%20Classification%20of%20SDG%20Indicators_13%20Dec%202023_web.pdf (accessed 10 December 2023).

Part III

The Human Context of Digital Work

4

Face to Face or Online: Towards a New Relational Governance of Building Design Projects

Jenni Barrett[1] *and James Sheppard*[2]

[1] *University of Central Lancashire, Preston, UK*
[2] *Equals Consulting, London, UK*

Introduction

In the latter years of the 20th century, and during the early part of the 21st century, architecture and construction have responded to emerging technologies that have improved efficacy and accuracy in the design and management of building projects. As computer-aided design technologies became more ubiquitous, the interdisciplinary architecture and construction team adapted their approaches to managing the collaborative design process. The emerging digital transfer of ideas and building information required a paradigm shift away from the traditional round table design meeting towards a new form which offered real-time sharing and data integrity. These incremental shifts in collaborative working were much discussed in the industry and explored by innovators who sought to embed the evolving communication technologies to improve collaboration. This was paralleled by a cultural shift in global work trends. Project teams were increasingly distributed, resulting from a shift towards the globalisation of communication and business internationalisation. Improved communication technologies also facilitated a new flexibility in work practice, away from the traditionally location dependent, simultaneously creating and responding to a growing demand for more flexible and autonomous working which could enable an equitable workplace and thus, more diverse teams. Organisations seeking to cast a wide net to attract the best talent, particularly those who were versant in the new digital skills, were keen to respond to that growing demand by embedding collaboration technologies in their work practices, so that they could offer more flexible careers and new forms of work contract. In parallel, shifts to more restrictive international work visa arrangements, rising fuel prices and intensifying global concern in relation to business travel's contribution to climate change; and the shift towards the distributed, networked team, that collaborates both synchronously and asynchronously towards project goals have gained significant momentum. In 2020, this shift in

Architectural Design and Management in the Digital Age: International Perspectives, First Edition.
Edited by Bob Giddings, Silvio Melhado, and Jenni Barrett.

the teamwork paradigm gained a new and unexpected imperative. The Covid-19 virus spread, requiring work teams to stay away from the office and avoid face-to-face contact with others. It became vital for those who were engaged in collaborative projects to find new ways of working together while still maintaining efficiency and achieving project success. Following the lockdowns that were experienced in many countries across the world, a new normal for working emerged. Before the Covid-19 pandemic, hybrid working featured only as a minority practice. Following the Covid-19 pandemic, hybrid working became the new normal. In 2022, the number of people in the UK who worked mostly from home and sometimes at their place of work continued to increase, from 30% in 2021 to 42% 2022. Those who planned to return to their place of work permanently continued to fall from 11% in 2021 to only 8% in 2022 (Office for National Statistics 2022). This trend is reflected across the world, although some regional differences exist, mostly due to nuances in national work cultures (Gensler Research Institute 2023). Before Covid, the new communication technologies offered novel, innovative solutions which teams might employ at their discretion. Following Covid, these technologies have become vital tools, critical to effective collaboration and project organisation. In the early 2020s, a global, digital and cultural transformation of work has enabled new perspectives, strategies and methods for the leadership and management of collaboration across built environment teams.

A New Relational Governance for the Digital Age

The ongoing cultural transformation in architecture and construction work necessitates a re-evaluation of communication systems and platforms available, and their scope for supporting collaboration strategies appropriate for industry and practice. The new normal for collaborative design and management of building projects benefits from the new and developing platforms for collaboration as they offer more efficient data and information exchange; real-time processing and improved integration and increased interoperability. Benefits of digital collaboration include efficiencies in project information management; enhanced workflow management; and communication for collaboration. These benefits certainly expedite and facilitate the sharing of information and instruction, but these techno-operational aspects describe only a fragment of the fuller definition of collaboration. As Larsson (2003: 159) pointed out, there is much more to design teams than merely exchanging data:

> *The essence of true collaboration is that team members are actually "thinking together" rather than only exchanging information and opinions. Meaning is collaboratively constructed in the negotiations, discussions and arguments that designers continuously have in their everyday work.*

The essence of human interaction in collaboration emphasises the fact that better connectivity does not necessarily mean better collaboration. As the latter in architecture, its design and construction may take place within digital or hybrid spaces, it is essential that people still control the contemporary and emerging technological environment. The concept of trust in collaborative relationships is perhaps the most commonly referenced as a crucial factor in creativity, innovation and the quality of collaboration. It is highlighted as a vital aspect in publications that were influential in industry culture change in the 20th century, such as the Latham's (1994) and Egan's (1998) reports, and underpinned new forms of contractual arrangements, such as partnering that sought to enhance industry performance through improved inter-organisational, interdisciplinary and supply chain relationships. Trust in collaborative relationships can be defined as respect without fear of criticism, and its presence is required for mutual reciprocation to be maintained and for innovative behaviour to occur. It can occur at an organisational level where a long-term collaborative relationship exists between two or more companies. In addition to inter-organisational trust, individual trust between participants in the design and construction process must also be present. Where an inclusive, socially cohesive group dynamic exists, a psychologically safe environment is created which is more likely to promote creativity and successful decision-making. This psychological safety is therefore considered a crucial antecedent to creativity and innovation, as a considerable amount of risk-taking, experimentation – and frequent failure – is required. Thus, team members are more likely to contribute ideas and solutions, if there is a high level of trust and a shared understanding in the team environment as a safe place for risk-taking. Where such an environment is not created, project participants are less likely to feel able to contribute, limiting the potential success of the collaborative relationship. In traditional project environments, cultures of psychological safety can be developed through social interactions during and around face-to-face intercommunication, particularly those that offer opportunities for informal connection and the development of a shared understanding of goals, motivation and expectations. Yet, in online and cyber-physical environments, geographical dispersal and the technological interface itself can act as a barrier to opportunities for this socio-psychological interaction and connection. Consequently, a lack of face-to-face interaction can be found to decrease the level of trust and shared understanding between project participants (Nguyen and Mougenot 2022). Given the importance of trust in collaboration, digitally enabled projects will require a conscious management of social relationships. This suggests that a relational, not just information or process governance approach is required, one where not only design and management protocols are implemented, but also where informality and personal connection can be nurtured.

A relational governance approach that supports trust in collaborative relationships can be accommodated in hybrid project environments. Where opportunities for face-to-face interaction present themselves, the agenda for these events should allow for, and even prioritise, informal interaction and connection through non-task-based discussion and socially led contact. Where online environments are the sole method of interaction, creating a climate of trust is more challenging. However, the technology itself can be used as an enabler of the relational governance strategy. For example, utilising appropriate platforms, depending on the nature of interaction or outcomes required, can form a significant part of relational governance. More sophisticated technologies such as blockchain that foster safe, secure and trustworthy information management are also increasingly being implemented in the construction industry as a mechanism to embed digital trust amongst collaborative teams (Mubarak and Petraite 2020). Relational governance can generate opportunities for the development of cultures of trust and psychological safety but it does not necessarily guarantee that such cultures will be developed or be maintained within the team. The strength, or weakness, of the social ties that subsequently develop between members are fundamental to the team's cohesiveness, and the degree of cohesiveness influences collaborative project performance. Team cohesiveness, as an aspect of team climate and dynamics, can be defined as the degree to which members align project goals and processes towards development of a shared understanding of commitment to the project. It describes the strength of the social ties between team members and is typically understood to be a precursor to successful collaborative performance. Cohesiveness, which is influenced by the degree of trust between team members, can be supported by the strengthening of social ties between those members and promotes good collaboration (Nell and Andersson 2012). These social ties are strengthened via social interaction, where connections between individuals or groups facilitate a sharing of information, knowledge and feelings about the task or team environment. Strong social ties and a high level of team cohesiveness are therefore typically considered a positive and aspirational attribute of the collaborative team, with high performance likely to be the outcome (Lokshin et al. 2009). This concept is an aspect of social network theory that conceptualises group members as nodes and implements measures to examine the centrality, grouping and closeness of these nodes to generate a visual representation of the cohesiveness of the network. Patterns of network cohesion thus indicate the presence of subgroups and unhelpful dispersion between group members as shown by the groupings, colour coding and distribution of nodes in Figure 4.1.

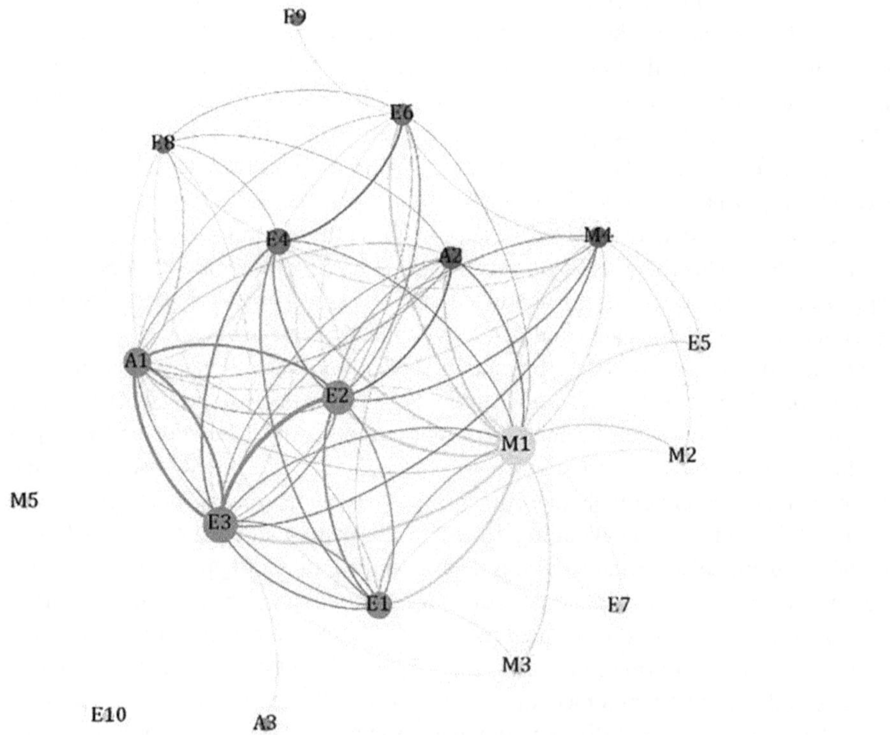

Figure 4.1 Social network pattern of online interactions of an observed building design team (Barrett 2019).

Intuitively, it might be expected that cohesive patterns with fewer subgroups and less dispersal would be more likely to perform better and be more creative in their output. However, this is not really the main issue in design and construction teams. Cohesion which influences high performance is dependent upon team longevity. This is rarely a feature of multidisciplinary teams as individuals tend to be brought together for the life of the project and disbanded or redeployed once the project or project phase is complete. It takes time for social relationships in teams to grow, and therefore social ties and cohesiveness may only characterise the team during the later delivery phases of a project. This presents a challenge for the early stages of design when collective goal setting and innovation are particularly required, but when social ties are likely to be at their weakest. Hallstrom et al. (2021) suggested that consideration of a threefold approach to social ties early in the project conception can help to ensure success in the early phases:

1) Selection and insight into the effects of a contract that shapes social bonds suitable for the nature of the collaborative network and communication required for the project (a formal state-type tie)
2) Knowledge sharing (informal event-type tie)
3) Trust (informal state-type tie).

These not only apply to traditional, face-to-face collaborative networks but also to digital and hybrid environments, where relational governance must embed formal and informal states and events to ensure success. Despite longevity being a factor in promoting cohesiveness and in turn good collaboration, it can also generate a familiarity amongst members who counter the good performance and collaboration that cohesiveness would typically facilitate. The familiarity of a very cohesive team may, over a longer time period, limit creative thinking as certain activities and interactions become routine, preventing the team from thinking outside its usual approach (Muskat et al. 2022). The digital, geographically dispersed environment may even help to limit this familiarity which, if governed effectively, can find the sweet spot for cohesive teams and promote high-performing collaborative networks.

Relational Governance and BIM

As the construction process for some projects has increased in complexity (Alshawi and Ingirige 2003; Chan et al. 2004), interdisciplinary architecture and construction teams have turned to building information modelling (BIM) to collaborate digitally. The benefits of enhanced virtual prototyping enabled by BIM have helped to reduce error, rework and waste, and in some instances, provide more sustainable design and construction (Khosrowshahi and Arayici 2012). BIM is a popular term and is no longer new in construction circles, having developed quickly since the millennium (Sun et al. 2017). Although BIM is commonly understood as a computer-aided drafting (CAD) derivative, it has no widely accepted precise definition. Sun et al. (2017) referred to it as a process of modelling, collaboration and integration while it has also been described as a italics multifunctional set of instrumentalities for specific purposes that will increasingly be integrated (Miettinen and Paavola 2014: 86It is a methodology comprised of technological, agential and managerial components (Oraee et al. 2017). Miettinen (2013) described BIM as a trans-discursive term in that it operates within research, policy and industry simultaneously. This form of digital technology encompasses more than the simple representation of a building, and it is akin to a socio-technical system (Liu et al. 2017), with true BIM sitting at the nexus of Culture, Process and Technology (Macdonald 2011). It is the lens that brings the requirement to collaborate into sharp focus. Oraee et al. (2019: 840) described collaboration as the italics 'agreement among several specialists to share their capabilities, including available data, information and knowledge in completing particular tasks, with the aim of achieving the project's broader objectives. Alreshidi et al. (2018) argued that collaboration precedents broadly fall within two categories, i.e. technical and socio-organizational factors. A superficial view would be that BIM is the technical pillar, but the literature sees it as an overarching process with italics software tools that are used simultaneously with non BIM tools (Miettinen and Paavola 2014: 87). There can be a tendency to view it simplistically as an integrated digital model with multi-disciplinary design inputs. However, it is more accurately described as a socio-technical process encompassing both software tools and an organising methodology to simultaneously direct design development alongside the model (Miettinen and Paavola 2014). Therefore, discussion around digital processes spans the creation of the 3-dimensional model, data-sharing portals, and online collaboration and meeting software. Designing with BIM involves developing an appropriate process, technology and culture (Macdonald 2011). Successful projects rely on effective social collaboration amongst BIM actors (Raja Mohd Noor et al. 2021). These actors form what can be referred to as BIM-based Construction Networks (Mignone et al. 2016), and their success relies upon the crucial role of collaboratively sharing project information across disciplines and different organisations (Merschbrock 2012; Oraee et al. 2019). They can be described as virtual teams, as their linkages and interactions are galvanised by the use of technology to communicate (Mignone et al. 2016).

There are many previous studies exploring why BIM uptake has been slower than anticipated; with the influencing factors categorised as technology, cost, management, personnel and legal aspects (Sun et al. 2017). Aside from the technology pillar, most are human, cultural or social issues. If the purpose is to promote collaboration (Sun et al. 2017), then the factors preventing the smooth deployment of BIM as a process can be viewed as collaboration blockers. Thus, there has not been sufficient need identified to disrupt the existing ways of working in favour of what may be perceived as more collaborative processes (Mignone et al. 2016). Oraee et al. (2019) identified prefer present tense collaboration in BIM-based Construction Networks as a multifaceted complex phenomenon affected by a variety of factors. They note the antecedents as context, team, process, task and actor, establishing that although the magnitude of academic research into barriers has increased since 2008, the proportion that is devoted to technical barriers dominates. While it has been established that BIM is a socio-technical system (Liu et al. 2017), the body of research suggests that comparatively scant attention has been focused on socio-cultural-related aspects of collaboration (Oraee et al. 2017). On the contrary, immature technology has been seen as the predominant barrier to collaboration, with attention paid to problems in technical development in terms of interoperability, easy data exchange, standardisation and virtual working (Idi and Khaidzir 2018; Oraee et al. 2019). In addition, the production of design information within a BIM model is the explicit product of a series of tacit decisions, which are rarely recorded or held anywhere other than in the brains of the designers (Dossick and Neff 2011). Therefore, the inseparable explicit and tacit knowledge is embodied and understood in the BIM model, solely by those who created it. This situation leaves *little room for the interpretive flexibility required for crossing inter organisational boundaries* (Neff et al. 2010: 564). So, while BIM is useful for explicit documentation and discovery of problems, it inhibits the ability to generate, share and informally exchange tacit knowledge (Dossick and Neff 2011). Having established the gap in research knowledge into collaboration amongst BIM-based Networks, the time has come to move beyond technology-oriented debate and instead focus on social interactions. Such attempts need established theories and a multidisciplinary approach to be applied in a construction context. Trust between parties is intrinsic to team members for successfully integrating relevant knowledge, expertise and skills (Liu et al. 2017). There is untapped potential to capitalise on virtual team working knowledge within Networks (Mignone et al. 2016), brought into sharp focus by the wider acceptance of virtual team working as a result of the COVID-19 pandemic. However, the increasing digitalisation within Industry 4.0 (Šuman et al. 2021) requires redefinition of the psychological contract in order to address satisfaction, engagement and other human relation issues (Coetzee and Deas 2021). Sutera et al. (2014) observed that the ability to collaborate is stifled by the erosion of psychosocial relations, brought about by virtual engagement together with incomplete and inadequate shared information. Tauriainen et al. (2016) concluded that the solution to Network collaboration problems does not require special hardware tools, instead citing awareness and improved design management practices as significant mitigators. Their recommendations include ensuring project managers are cognisant of BIM as process, or at least supported by someone who is knowledgeable about it. Within BIM governance, Alreshidi et al. (2017) identified the importance of team and actor characteristics to effective collaboration, as well as technical and data elements. It is argued that although BIM supports the exchange of knowledge, this exchange is explicit and not flexible or informal enough to exchange the tacit knowledge exchanged through just talking (Dossick and Neff 2011).

Future tools to support collaboration need to blend the experience of both deductive embodied experiences from the inexperienced user and also the efficiency of the expert user's immediate solution in order to promote creativity (Sutera et al. 2014). Regardless of technical experience, users need to feel that they are in a culture of openness and trust, where they are safe to express their feelings to feed into decision-making processes (Blundell 2021). This behaviour needs to be encouraged and nurtured without the harness of stifling contractual, cultural and organisational boundaries (Dossick and Neff 2011). Real-time creation and the sharing of information between parties enable discussion. It requires flexibility, action and informality in addition to mutual discovery, critical engagement and knowledge exchange (Dossick 2014). Despite being well established as a socio-technical process, the evidence suggests that there is a tendency to oversimplify the use of BIM as a suite of digitally enabled tools and underplay the importance of the inherent Network collaboration process, which is currently limited by implied organisational barriers (Liu et al. 2017). This is exemplified by the fact that attempts are mostly made to address these issues with technology (Idi and Khaidzir 2018; Oraee et al. 2019). It becomes evident that project managers should pay attention to the use of technology to adjust organisational structures, when dealing with interpersonal issues (Mignone et al. 2016). BIM-based Project Management (Ma et al. 2018) suggests a new paradigm to integrate project management into functional BIM applications to promote collaboration. Oraee et al. (2019) suggested prefer present tense that those wishing to project manage BIM-based Networks to promote collaboration should understand both technological and interpersonal factors regarding collaboration and concentrate on developing relationships when managing virtual teams. Ultimately, as the dominance of technical push recedes and softer skills come into prominence, specialist BIM roles and responsibilities will be incorporated by project managers (Hosseini et al. 2018). Dossick and Neff (2011) identified prefer present tens that there is a gap in the knowledge of what is required in management and leadership decisions required to foster collaboration. The issue that comes to the fore is how to unlock true collaboration and move beyond information exchange that solves identified problems, to something more akin to people thinking together (Larsson 2003).

Towards a Framework for Collaboration in BIM Practice

In 2022, research was conducted within the UK construction industry to establish the scope of BIM processes as a socio-technical collaboration approach (Sheppard and Barrett 2022). As creativity and innovation are identified as social outcomes, i.e. those produced by human interaction, the research also examined the extent to which BIM processes enable or block creativity in collaboration. The objectives of the research were to

- establish a practitioner-derived view of how effectively BIM processes enable creative collaboration within the UK construction industry.
- suggest a future framework to facilitate creative collaboration in BIM processes, and to gauge how amenable the industry could be to this proposal.

The research was carried out using a mixed-method approach incorporating grey and academic literature reviews, practitioner interviews and surveys. The mixed method was chosen to triangulate and test the emerging themes. The practitioner interview sample pool was drawn from design practitioners, within the UK construction industry, who are actively involved in digital design management processes to an experienced level. The interview data from 255 industry practitioners were coded and thematically reviewed and used to inform the emerging framework. A focus for the research was the BIM Execution Plan, a structured collaborative planning guide (Anderson and Ramalingam 2021), which is considered necessary for the success of a project. Therefore, it was important to establish how real-world execution plan examples address creative collaboration and to what extent they can define the circumstances for it to flourish. Several interpretations of the word collaboration were reviewed and featured significantly within all the sample documents; either incorporated into a title or forming part of a BIM objective. However, it was often solely associated with the technical requirements of Quality Assurance or the establishment of Common Data Environment formats, rather than people-focussed elements. When considering the many factors that contribute to collaboration, this is limiting, especially from the social perspective. Nevertheless, the theme of digital BIM processes as an enabler of social and creative collaboration was strongly endorsed during the interviews, with practitioners recognising that BIM can aid communication by creating common ground, as it can be more efficient and democratic. For instance, younger, less-experienced colleagues can gain beneficial experience of exchanges that that they would not have been able to experience previously, due to time and travel pressures. Some participants also noted that it mitigated against inherent gender barriers within a male-dominated industry, for example,

> *So, it has reduced a status difference between junior people and senior people... removing the... height difference that might create a dominance situation between male and female. Those things reduce the entry barriers to people participating in a meeting. (Participant G)*

Similarly, survey respondents overwhelmingly felt that BIM processes enabled collaboration, with 79% tending to agree and only 3% disagreeing. However, as can be seen from Figure 4.2, when comparing this to the extent to which they feel the collaboration is creative, this drops to only 34% tending to agreement, with 25% disagreeing and 41% unsure.

Observations drawn from the interviews are that digital processes block creative collaboration because they are limited, rigid and tend towards transactional exchanges. There is a feeling that the federated digital model has reached formal maturity and therefore difficult or costly to retract previous design decisions. One participant observed,

> *Now, you go straight into BIM... That all seems premature to me. It all seems far more resolved or has the illusion of being far more resolved. So... we're taking advantage of the digital tools, but there's no sort of reciprocal flexibility that we're offering to recognise that it's early in the scheme: we want all that information and then we don't want it to change. (Participant L)*

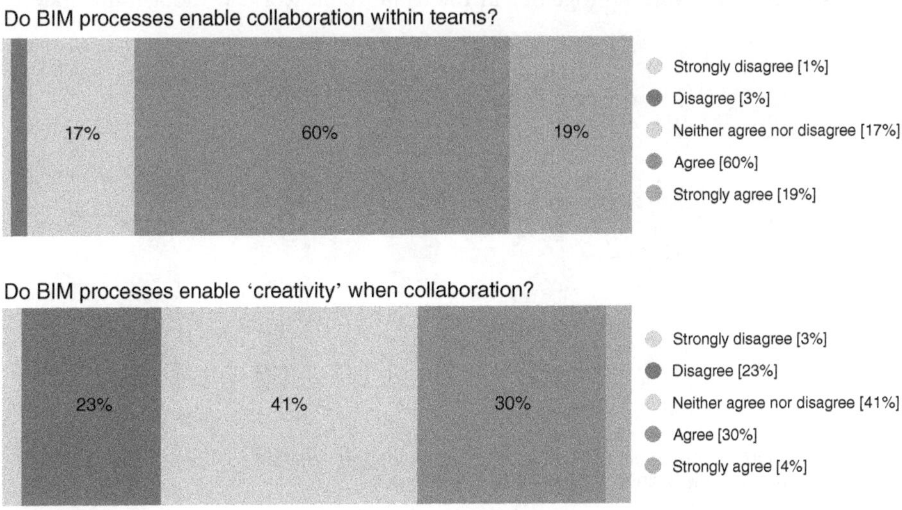

Figure 4.2 Extent that BIM processes enable creative collaboration (Sheppard and Barrett, 2022 / University of Cambridge).

Participants felt that information presented online is at a more complete stage, without sharing the working out processes that informed it, which are arguably more suited to creative input from other disciplines, as noted by another participant:

> *You miss out on a huge amount of fuzzy stuff which has happened . . . but those dead ends and things, they're not wrong. . . . They were an important part of the process at the moment. (Participant K)*

Online meetings are often one party demonstrating to others, rather than collaborative collective sharing of a creative environment. There is a transactional format with either stated or implied agendas, where design attention is often limited to resolving clash detection lists, to the exclusion of wider creative process. There is no time for hypotheticals or sharing the working out of scenarios, as well as a lack of opportunity to develop personal relationships with others. Two participants expressed frustration with this rigidity:

> *. . . whereas it could have been something that you just pick up the phone to someone on the side and say in an informal way, they just sort it out. And that's what's quite tricky, I think, in the digital meeting format, kind of project team meetings, because it's very kind of rigid in a way. (Participant J)*

> *When we went virtual, something was lost in terms of the incidental. Everything became formal. (Participant L)*

When asked to rank where creative collaboration is most likely within a selection of digital BIM processes, there was a strong preference for some form of contact between people, as can be seen in Figure 4.3.

When the interviewees were questioned on what enabled creative collaboration, many participants mentioned working together in the same room work, without immediately

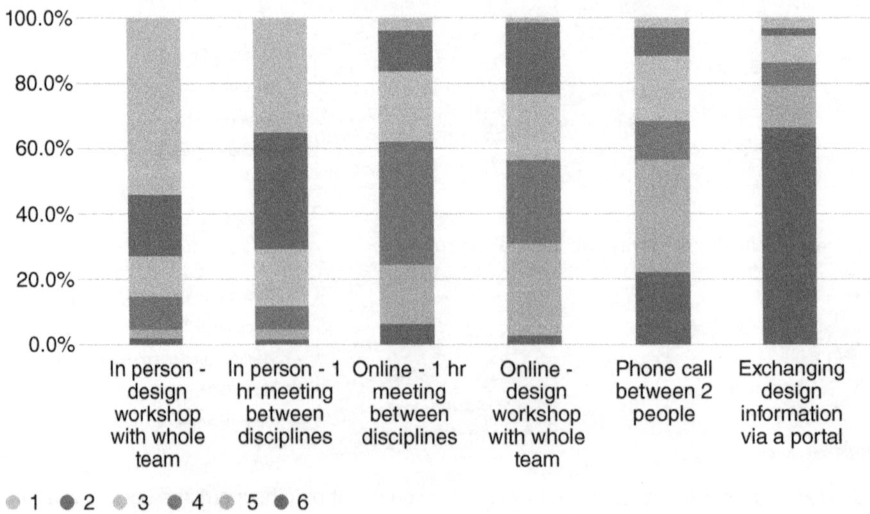

Figure 4.3 Circumstances where creative collaboration is more likely to occur during BIM processes (Sheppard and Barrett, 2022 / University of Cambridge).

being able to articulate what behaviour this promoted. Upon further discussion, the following themes emerged:

1) trust, to share own ideas and be open to those from others
2) forming relationships
3) enjoying the process
4) being open minded and willing to engage.

Social interaction contributes to the development of trust and motivation and 79% of participants tended to agree that social interaction increased the likelihood of creative collaboration, as can be seen in Figure 4.4.

However, when asked whether good-quality social interaction occurred specifically during digital BIM processes 47% tended to disagree, with 37% undecided, as can be seen in Figure 4.5.

Motivation and enthusiasm are also affected by levels of fulfilment and when asked about this issue, only 16% of respondents felt socially fulfilled during BIM processes compared to 58% being professionally fulfilled, as shown in Figure 4.6.

The importance of meeting in person to effect creative collaboration is more than simply reading visual cues; it also helps form social relationships. The incidental conversations that occur in and around the formal agenda help to form common bonds between project participants. One participant linked the loss of this situation to the online meeting process:

> *You don't have a chat on the way out . . . there is something lost in that. And also, you're back-to-back in a (MS) Teams World . . . you don't really you have the bandwidth for that kind of effort, to make the effort around that. . . . putting effort into sort of breaking down some of those barriers, I guess, as part of a design process, arguably needs recognizing and focus more because it doesn't happen ad hoc. Maybe you need to make time for it. (Participant L)*

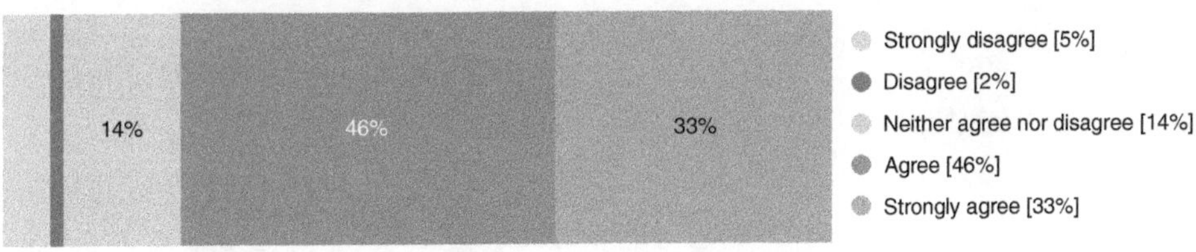

Figure 4.4 Social interaction increasing the likelihood of creative collaboration (Sheppard and Barrett, 2022 / University of Cambridge).

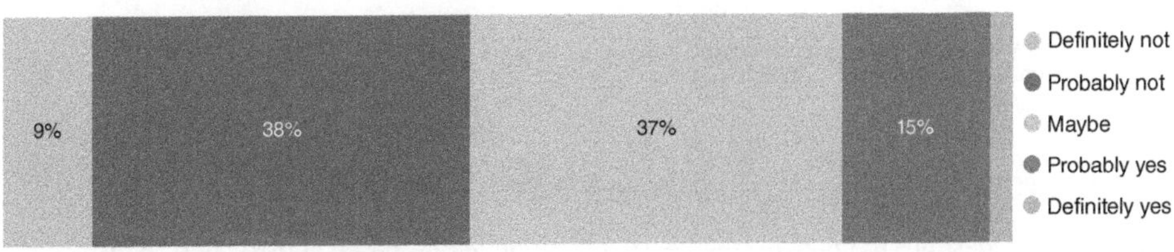

Figure 4.5 Good-quality social interaction occurring during online BIM processes (Sheppard and Barrett, 2022 / University of Cambridge).

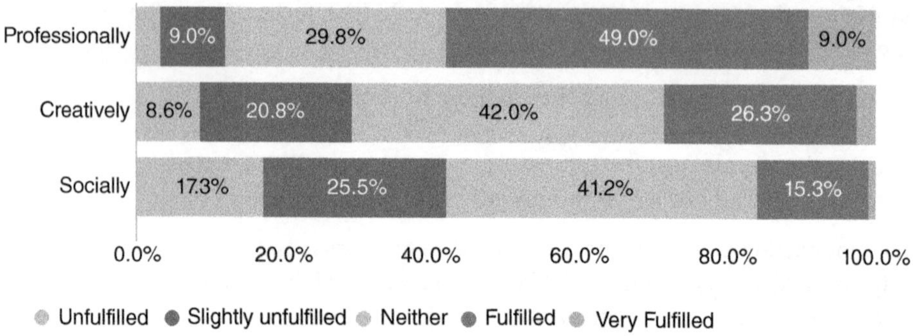

Figure 4.6 Fulfilment perception when interacting with others in BIM processes (Sheppard and Barrett, 2022 / University of Cambridge).

Given that there is a natural wariness of new technology, it is not surprising that there is an underlying desire to return to traditional design processes. It would be easy to dismiss this as Luddite simplicity without delving into more detail. There could be a genuine impactful loss as a side effect of the efficiencies of rigid digital processes. By reducing social contact, it could be that some of the circumstances required to promote creative collaboration are removed. Building on the thematic observations, the survey included some questions around possible future techniques to shape thinking of what might be done to improve processes. When asked whether a BIM execution plan would be improved in terms of enabling creative collaboration and if it also touched on human social aspects, 72% tended to agree with only 12% tending to disagree. Issuing data through a portal is ranked as the least likely to promote creative collaboration, so mutually agreeing to always talking through the deliverable with the receiving party was felt to be a welcome improvement for 84% of respondents (see Figure 4.7). In addition, 67% of respondents also agreed that BIM processes should have social opportunities built into the process.

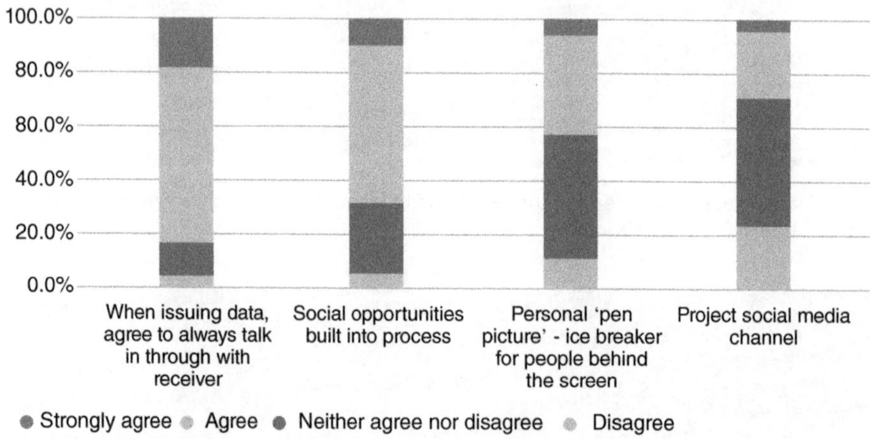

Figure 4.7 Suggestions to improve creative collaboration in BIM-based Construction Networks (Sheppard and Barrett, 2022 / University of Cambridge).

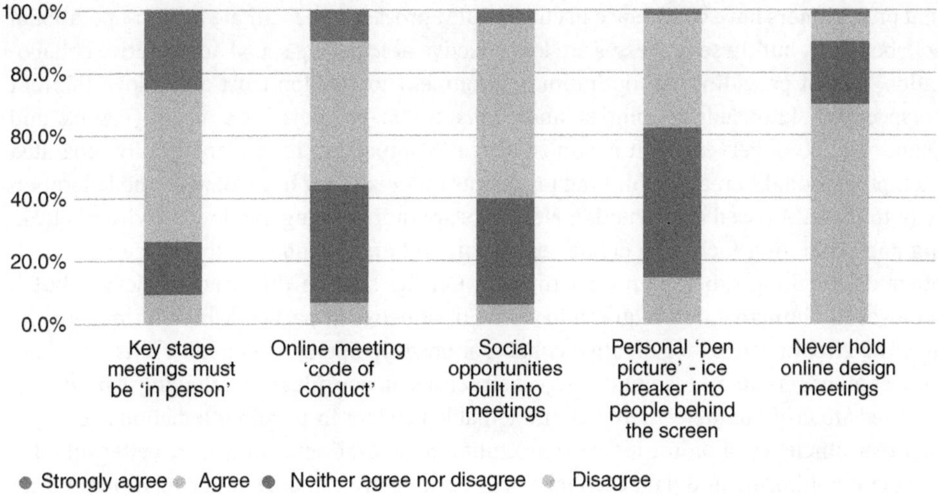

Figure 4.8 Features that would improve creative collaboration in meetings (Sheppard and Barrett, 2022 / University of Cambridge).

Respondents were asked their opinion on aspects to improve creative collaboration in meetings. Referring to Figure 4.8, the usefulness of online meetings is undeniable, with 70% of respondents feeling that it would be detrimental to remove them. Some suggestions for improvement are building in social opportunities (60% tended to agree), personal pen picture or profile as an ice breaker (38% agreed) and mutually agreeing an online meeting code of conduct (56% agreed). However, 74% agree that key stage meetings should always be in person, reinforcing the importance for a mixed-mode strategy being considered when strategising a design process.

Conclusion

The digital transformation of the construction industry and the cultural shift in work in the 21st century now requires a rethink of the paradigms of teamwork and collaboration. Digital platforms and processes have delivered many efficiencies and benefits that have enabled significant technical and operational successes, yet the effect on people has not been aligned. Without this consideration, true collaboration is limited, and in turn the potential for innovation in wider design success. A new project relational governance paradigm for the digital age is essential, if BIM is to be pursued, and it needs to be determined if people can feel comfortable within the new technological environment. Architecture and construction disciplines are urged to perpetuate social processes, not as an optional beneficial process innovation, but with at least equal priority to digital and technical solutions. Design strategies must similarly intertwine the social factors with operational goals, if high-performing teams can deliver high-quality design outcomes for owners, users and society. As a global, widely used, and sometimes a mandatory digital process, BIM is examined for its capacity to facilitate effective and creative collaboration. The findings indicate that practitioners have confidence in current BIM processes to facilitate techno-operational collaboration, but these processes are less effective at fostering social and creative collaboration; by not providing the optimum environment to develop trust and allow different perspectives, knowledge, abilities and skills to be employed on tasks (Paulus and Dzindolet 2008). Perversely, it is almost as if BIM processes have ideologically separated each professional's creative thinking process as much as they have allowed the federation of outputs. BIM-based Construction Networks are more willing to allow the virtual cloaking and separating factors to create separation and are not obliged to engage with each other on anything other than data transfer. On the surface this seems efficient, but it removes the human spark required to create trust between parties, which ultimately creates the circumstances for creative collaboration to happen. To negotiate this situation, project managers should facilitate project structures to consider the following aspects:

Mixed Mode of Contact – there is an undeniable need for in-person interaction to compliment the efficiency of online interaction. Online collaboration meetings are better suited to progress monitoring and data exchange once teams have established the environment for trust, social ties, and team cohesiveness to develop. In-person workshops at key stages can be viewed as a necessary way of opening communication from beyond the increasingly digital veil of siloed data transfer.

Behaviour – the requirement to attend certain meetings in person should be codified into engagement contracts. Moving further, the project manager should communicate why this is important. Engagement could be shaped to include some of the suggestions above. Participants need to agree to buy into beneficial behaviour even when engaging online, such as agreeing to call someone and explain the data uploaded to a portal, commitment to not carrying out emails or other work tasks during meetings, cameras turned on, etc.

BIM Execution Plan Format – This should be reviewed to provide more emphasis on human collaboration, as described above, to establish a truly universal Common Data Environment.

While these findings emerge from research that focuses on BIM application, they are applicable to all digitally enabled interdisciplinary building projects. Projects that can embed a relational governance strategy that formalises within it, these necessary modes of contact and behaviour are more likely to facilitate better collaboration and thus, better building design outcomes and performance. Researchers and practitioners are invited to develop and test these concepts, so that a detailed and universal framework for the socio-organisational dimension of design can be realised and applied. This is likely to support future technological maturity and industry evolution that can bring to bear an improved trajectory for the future of architecture in the digital age.

References

Alreshidi, E., Mourshed, M. and Rezgui, Y. (2017). Factors for effective BIM governance. *Journal of Building Engineering*, 10, 89–101. https://doi.org/10.1016/j.jobe.2017.02.006.

Alreshidi, E., Mourshed, M. and Rezgui, Y. (2018). Requirements for cloud-based BIM governance solutions to facilitate team collaboration in construction projects. *Requirements Engineering*, 23(1), 1–31. https://doi.org/10.1007/s00766-016-0254-6.

Alshawi, M. and Ingirige, B. (2003). Web-enabled project management: an emerging paradigm in construction. *Automation in Construction*, 12(4), 349–364. https://doi.org/10.1016/S0926-5805(03)00003-7.

Anderson, A. and Ramalingam, S. (2021). A socio-technical intervention in BIM projects – an experimental study in global virtual teams. *Journal of Information Technology in Construction*, 26, 489–504. https://doi.org/10.36680/j.itcon.2021.026.

Barrett, J. (2019). Spaces of collaboration: exploring social exchange in the digital domain. *World Building Congress*. Hong Kong: CIB.

Blundell, S. (2021, November 1). The human side of digitalisation in the built environment. *Planning, BIM & Construction Today*. https://www.pbctoday.co.uk/news/bim-news/digitalisation-in-the-built-environment/101266/

Chan, A.P.C., Scott, D. and Chan, A.P.L. (2004). Factors affecting the success of a construction project. *Journal of Construction Engineering and Management*, 130(1), 153–155. https://doi.org/10.1061/(ASCE)0733-9364(2004)130:1(153).

Coetzee, M. and Deas, A. (2021). *Redefining the Psychological Contract in the Digital Era: Issues for Research and Practice*. London: Springer Nature.

Dossick, C.S. (2014). Messy work in virtual worlds: exploring discovery and synthesis in virtual teams. In Y. Luo (ed.) *Cooperative Design, Visualization, and Engineering*. Cham: Springer International Publishing (Lecture Notes in Computer Science). 134–142. https://doi.org/10.1007/978-3-319-10831-5_19.

Dossick, C.S. and Neff, G. (2011). Messy talk and clean technology: communication, problem-solving and collaboration using building information modelling. *Engineering Project Organization Journal*, 1(2), 83–93. https://doi.org/10.1080/21573727.2011.569929.

Egan, J. (1998). *Rethinking Construction: Report of the Construction Task Force*. Norwich: HMSO.

Gensler Research Institute. (2023). *Global Workplace Survey Comparison*. San Francisco: Gensler.

Hallstrom, A., Bosch-Sijtsema, P., Polete, L., Rempling, R. and Karlsson, M. (2021). The role of social ties in collaborative project networks: a tale of two construction cases. *Construction Management and Economics*, 39(9), 723–738.

Hosseini, M.R., Martek, I., Papadonikolaki, E., Sheikhkhoshkar, M., Banihashemi, S. and Arashpour, M. (2018). Viability of the BIM manager enduring as a distinct role: association rule mining of job advertisements. *Journal of Construction Engineering and Management*, 144(9), 04018085. https://doi.org/10.1061/(ASCE)CO.1943-7862.0001542.

Idi, D.B. and Khaidzir, K.A.M. (2018). Critical perspective of design collaboration: a review. *Frontiers of Architectural Research*, 7(4), 544–560. https://doi.org/10.1016/j.foar.2018.10.002.

Khosrowshahi, F. and Arayici, Y. (2012). Roadmap for implementation of BIM in the UK construction industry. *Engineering, Construction and Architectural Management*, 19(6), 610–635. https://doi.org/10.1108/09699981211277531.

Larsson, A. (2003). Making sense of collaboration: the challenge of thinking together in global design teams. *Proceedings of the 2003 International ACM SIGGROUP Conference on Supporting Group Work*, 153–160. https://doi.org/10.1145/958160.958184.

Latham, S. (1994). *Constructing the Team: Final Report of the Government/Industry Review of Procurement and Contractual Arrangements in the UK Construction Industry*. London: HMSO.

Liu, Y., van Nederveen, S. and Hertogh, M. (2017). Understanding effects of BIM on collaborative design and construction: an empirical study in China. *International Journal of Project Management,* 35(4), 686–698. https://doi.org/10.1016/j.ijproman.2016.06.007.

Lokshin, B., Gils, A. and Bauer, E. (2009). Crafting firm competencies to improve innovative performance. *European Management Journal,* 27(3), 187–196.

Ma, X., Xiong, F., Olawumi, T., Dong, N. and Chan, A. (2018). Conceptual framework and roadmap approach for integrating BIM into lifecycle project management. *Journal of Management in Engineering,* 34, 1–10. https://doi.org/10.1061/(ASCE)ME.1943-5479.0000647.

Macdonald, J.A. (2011). *BIM – Adding Value by Assisting Collaboration.* 11.

Merschbrock, C. (2012). *Unorchestrated Symphony: The Case of Inter-Organizational Collaboration in Digital Construction Design.* 18.

Miettinen, R. (2013). *Innovation, Human Capabilities, and Democracy: Towards an Enabling Welfare State.* Oxford: OUP.

Miettinen, R. and Paavola, S. (2014). Beyond the BIM utopia: approaches to the development and implementation of building information modeling. *Automation in Construction,* 43, 84–91. https://doi.org/10.1016/j.autcon.2014.03.009.

Mignone, G., Hosseini, M.R., Chileshe, N. and Arashpour, M. (2016). Enhancing collaboration in BIM-based construction networks through organisational discontinuity theory: a case study of the new Royal Adelaide Hospital. *Architectural Engineering and Design Management,* 12(5), 333–352. https://doi.org/10.1080/17452007.2016.1169987.

Mubarak, M. and Petraite, M. (2020). Industry 4.0 technologies, digital trust and technological orientation: what matters in open innovation? *Technological Forecasting and Social Change,* 161, 120–332.

Muskat, B., Anand, A., Contessotto, C., Tan, A. and Park, G. (2022). Team familiarity—boon for routines, bane for innovation? A review and future research agenda. *Human Resource Management Review,* 32, 100892.

Neff, G., Fiore-Silfvast, B. and Dossick, C.S. (2010). A case study of the failure of digital communication to cross knowledge boundaries in virtual construction. *Information, Communication & Society,* 13(4), 556–573. https://doi.org/10.1080/13691181003645970.

Nell, P. and Andersson, U. (2012). The complexity of the business network context and its effect on subsidiary relational (over-) embeddedness. *International Business Review,* 21(6), 1087–1098.

Nguyen, M. and Mougenot, C. (2022). A systematic review of empirical studies on multidisciplinary design collaboration: findings, methods, and challenges. *Design Studies,* 81, 101–120.

Office for National Statistics. (2022, May 23). *Is Hybrid Working Here to Stay?* Retrieved 4 January, 2023, from Office for National Statistics. https://www.ons.gov.uk/employmentandlabourmarket/peopleinwork/employmentandemployeetypes/articles/ishybridworkingheretostay/2022-05-23 (accessed 4 January 2023).

Oraee, M., Hosseini, M.R., Papadonikolaki, E., Palliyaguru, R. and Arashpour, M. (2017). Collaboration in BIM-based construction networks: a bibliometric-qualitative literature review. *International Journal of Project Management,* 35(7), 1288–1301. https://doi.org/10.1016/j.ijproman.2017.07.001.

Oraee, M., Hosseini, M.R., Edwards, D.J., Li, H., Papadonikolaki, E. and Cao, D. (2019). Collaboration barriers in BIM-based construction networks: a conceptual model. *International Journal of Project Management,* 37(6), 839–854. https://doi.org/10.1016/j.ijproman.2019.05.004.

Paulus, P.B. and Dzindolet, M. (2008). Social influence, creativity and innovation. *Social Influence,* 3(4), 228–247. https://doi.org/10.1080/15534510802341082.

Raja Mohd Noor, R.N.H., Che Ibrahim, C.K.I. and Belayutham, S. (2021). Making sense of multi-actor social collaboration in building information modelling level 2 projects: A case in Malaysia. *Construction Economics and Building* 21(4), 89–114. https://doi.org/10.5130/AJCEB.v21i4.7829.

Sheppard, J. and Barrett, J. (2022). Building on binary: integrating creative collaboration within digital design processes in the construction industry. *Research Study* (*unpublished*). University of Cambridge.

Šuman, N., Pučko, Z., Šuman, N. and Pučko, Z. (2021). *Integration of Industry 4.0 for Advanced Construction Project Management.* IGI Global. https://services.igi-global.com/resolvedoi/resolve.aspx?doi=10.4018/978-1-7998-8548-1.ch065; https://www.igi-global.com/gateway/chapter/276876

Sun, C., Jiang, S., Skibniewski, M.J., Man, Q. and Shen, L. (2017). A literature review of the factors limiting the application of BIM in the construction industry. *Technological and Economic Development of Economy*, 23(5), 764–779. https://doi.org/10.3846/20294913.2015.1087071.

Sutera, J., Yang, M.C. and Elsen, C. (2014). The impact of expertise on the capture of sketched intentions: perspectives for remote cooperative design. In Y. Luo (ed.) *Cooperative Design, Visualization, and Engineering.* Vol. 8683. Springer International Publishing. 245–252). https://doi.org/10.1007/978-3-319-10831-5_36

Tauriainen, M., Marttinen, P., Dave, B. and Koskela, L. (2016). The effects of BIM and lean construction on design management practices. *Procedia Engineering,* 164, 567–574. https://doi.org/10.1016/j.proeng.2016.11.659.

5

Creativity in the Digital World

Maria Vogiatzaki[1], *Bob Giddings*[2], *and David Morton*[2]

[1] *Anglia Ruskin University, Chelmsford, UK*
[2] *Northumbria University, Newcastle upon Tyne, UK*

Introduction and Context

It is an established theme that architectural design is the act of transforming ideas into buildings (Brawne 1992). In this respect, tool invention, development and deployment have always been key challenges. The invention of a tool, as any cultural artefact, is affected by the context in which it unfolds (Colomina and Wigley 2016). The perspective, as a key drawing, developed in the Enlightenment as a hymn to the dominance and sovereignty of human discovery. Debate on the validity of images that represent architecture has a long history, with conflicting attitudes to the art of drawing. Issues of skill, honesty and artistic licence apply to all forms of architectural representation and should be considered in any analysis of correlation between architectural representation and eventual building (Stamp 1982). These kinds of concerns became particularly related to perspective drawing, and it began to lose favour (Giddings and Horne 2002). The prevalence of rationalism and the evolution of the sciences became embraced in the design process, prioritising function and spatial organisation explored through dissecting a building with plans, sections and elevations. This led to the desire for a more intellectual approach to architecture with tools such as statistics, qualitative and quantitative data and analysis. However, it also brought to the fore, concepts of participatory design and the involvement of disciplines such as sociologists and anthropologists. The prevailing cultural context in each time and place demands its priorities and values to be reflected and determine which tools should be employed.

A major turning point in the history of architectural design is the pervasiveness of information technology and the way it affects current practice. The translation of an idea to a building within an architectural practice was the privilege of its principal. There is now debate as to whether buildings can be created by a gifted individual or whether digitisation demands that all work has to be undertaken in teams, whether they are monodisciplinary, multidisciplinary or interdisciplinary. Digitisation has much closer association with building production, and with the development of building information management, an idea

Architectural Design and Management in the Digital Age: International Perspectives, First Edition.
Edited by Bob Giddings, Silvio Melhado, and Jenni Barrett.

can be directed or even generated from a database. It is argued that the nature of projects, especially at a grand scale and complexity, cannot be managed by human cognition alone. Computerisation aids remote operations at both design and building phases of a project, offers processing and combining project databases. It can even drive construction both remotely through software, or generating the actual building through automation and the deployment of digitally fabricated, 3d printing (Carpo 2011). In this context, where architectural designers are increasingly required to translate architectural ideas into building data and design within a database, there are concerns about how creativity can thrive. Moreover, it implies a kind of standardised building project. Of all industries, the products of the construction industry are the most varied. The size and scale of a proposal, intended uses, location, surroundings, site, community, materials and construction skills – are just some of the variables that make each building design unique.

This chapter will reflect on this conundrum and upon the role of the architect in the digital world. It will explore creativity in the context of a digital project, with specific attention to the nature and value analogue techniques such as of drawing and modelling. This is often presented as a binary divide, and it may be that the most stimulating creativity can be generated by combinations and integration of analogue and digital methods at different stages of the design process. The notion of digital creative authorship will be raised, questioning the status of 'starchitects' in an industry increasingly embraced by the shared authorship of digital models, issues of intellectual property and blurring professional roles in the 21st century project teams. This chapter will also consider the challenges to creative thinking presented by remote working and will discuss the potential for creativity without the social rewards and motivational environment offered by the office or studio environment.

The Concept of Creativity in Architectural Design

In simple terms, creativity is a process from which an intangible or physical object is formed, where it did not previously exist. Creativity can be found in a number of disciplines, but is most associated with the arts – 2d and 3d design, literature, visual and performing arts, music and architecture. Bartley (2023) referred to it as new ways of inhabiting the world. To be truly creative, the products must be useful. Yet, this usefulness may be difficult to measure. In many cases, it might only be achievable through vicarious means. For example, the objective of music is to enhance the emotions, but its success is measured by how many people are willing to purchase it. Yet, it is debatable whether there is a correlation between commercial gain and true creativity. There are aspects of architecture that are measurable. These can be termed performance attributes and especially fitness for purpose. They include – building utilisation, temperature/ventilation/illuminance/sound, energy utilisation, operational efficiency, buildability and user satisfaction. Nevertheless, to be perceived as architecture, amenity attributes must be considered. These involve atmosphere, character, image, symbolic significance, contextual impact and aesthetics, which are not so easy to measure. According to Resnick (2017), it is not true that creative people are a rarity, with special characteristics of being smart, curious and able to look at the world with fresh eyes. Indeed, there are a number of misconceptions, including that creativity cannot be taught. According to Boden (1998), there are three main types of creativity: exploratory, transformational and combinational. The first type is the process of searching a conceptual space that is already governed by specific rules and boundaries. It is often perceived in improvements to established objects, and where iteration can be enacted. It would seem that this type is particularly applicable to digitisation. Transformational creativity is deployed to make the world a better place, for example, composing a piece of music that is not to be exploited commercially. However, Bartley (2023) noted that to translate creativity into reality, it takes ability. It charges creativity with novelty, uniqueness, relevance and reason. This approach does not seem to fit systems very well and may be associated with individuals such as mavericks. It certainly takes more than just pressing the buttons on a machine. A combinational approach is the investigation of unfamiliar combinations, and the process of bringing together old ideas to generate something else. In a world of great ideas from the past, there is a debate about how new any idea can be, and whether creativity is about novelty. North (2013) discussed the possible paradox of how original an idea might be when it builds upon existing ones and whether it really matters. This could relate to a series of architectural designs, such as the Hampshire Schools and Arup's commercial developments in the United Kingdom, from the late 20th century. Design is also linguistically conditioned, and issues of vocabulary are much underestimated.

An inhibition to creativity may be attributed to language and perception. It was pointed out by Austrian-British philosopher Ludwig Wittgenstein (1889–1951) that the nuances of language can limit interaction (Lockey 2023). Contemporary materialist philosophers, such as DeLanda (2013) also attributed prefer present tense the inhibition of creating the new to culturally conditioned language. A built environment that is designed by one set of professionals, constructed by another group, inhabited by others and demolished or repurposed by a different set makes architecture the most public of arts that cannot be socially autonomous and distanced from its social function (Drummond 2023). Thus, in architecture, an idea cannot be completely limited to one person's private sphere and needs to be shared, validated and accepted by professionals, users and indeed society. In order to achieve this consensus, architectural debate requires a common language and vocabulary, and the same way of 'seeing' the world. In using digital systems, architects invest in developing the sophisticated languages of toolkits, which establishes a kind of authority for their role in shaping the built environment. It also means a type of exclusion for those who are not privy to this vocabulary. They may become less able to communicate and be involved in co-creation, limiting accessible for all, until the design is presented. Ironically, terms such as intelligent environments, interfaces and user friendliness are used to describe accessibility to the innovation of incomprehensible computer programs. At the other end of the spectrum, graphic programs that facilitate simplistic visual connections miss the breadth and complexity inherent in creativity.

Current value systems indicate that interventions in the built environment are principally shaped by two opposing tendencies. The first is about monumentality, bigness and signature. The second one stems from community, context, climate change and socio-political priorities around the globe. These concerns have spread around architectural professionals in the last decade. Nonetheless, monumentality and iconic, signature architecture continue to thrive in the global consumerist world (Nicolai 2013). These projects have no specific geography and continue to prosper as a result of consensual neo-liberal socioeconomic policies, transnational networking of companies and individuals, flows of global finance and rapid restructuring of labour markets. This has given rise to emblematic projects such as Bilbao's Guggenheim, Grand Paris, Bruxelles 2040 and Dubai's Burj Khalifa and Louvre (Ginichiu and Schoenmaeckers 2018). Iconic buildings are integral to and symbolic of power. They are reminiscent of the architecture of Speer and Terragni for Hitler and Mussolini, respectively, to the hymn of the industrial revolution and unprecedented high-rise buildings through artful high-tech engineering and material innovation (Sklair 2012). There can also be attempts to camouflage these power systems by apparent sustainable strategies.

Koolhaas's approach to the 'Common Grounds' Biennale, curated in 2014, was to anathematise the monumentality of architecture. Despite showcasing the work of anonymous architects from the Netherlands, France, Italy, the United Kingdom and Germany for their contribution to community, not eroded by the globalised consumerism, reality suggests this approach is naively optimistic. Almost 10 years of reflection and further contribution of the built environment to the climate crisis, monumentalism is losing ground. In this world, in which flamboyance and style have long determined how an architect becomes a star, the low-tech, low-profile socially sensitive, but above all unknown to date, Pritzker laureates of the last two years is an institutional shift to what could be termed a Gaia-centred approach to architecture (Vogiatzaki 2018). With their philosophy of aiming to refurbish, rather than rebuild social housing, Lacaton and Vassal were awarded the built environment's equivalent of the Nobel Prize is a revolution. As the jury puts it, Lacaton and Vassal have not only renewed the legacy of modernism, but they are also redefining architecture itself (Vogiatzaki 2021). The 2022 Pritzker Laureate stated that he is hoping to change the paradigm and enable people to dream. He said that the poor should not be deprived of quality, as everybody deserves quality, comfort and luxury (Kéré 2022). This does not only represent an institutional change but also suggest a new and revised built environment with projects whose aesthetics are shaped by ethics and its politics. Therefore, while novelty may be a driving force behind the work of starchitects, it might not be as significant to building users as providing appropriate environments.

Changing Practices

By its nature, current building design processes entail relationships with others, although these interactions vary greatly with the type of project and the circumstances in which they are conceived. There has always been change in the way that people work, but it is argued that it has accelerated in recent years. Up to the middle of the 20th century, the image was one of the individual architects demonstrating a commanding position in architectural design. Nobody is mentioned in the design of Blenheim Palace, except for Vanbrugh. In the 19th century, George Gilbert Scott had the biggest practice that any one architect had ever enjoyed in Britain. He became the president of the Royal Institute of British Architects, Royal Gold Medallist, professor of architecture at the Royal Academy and was widely regarded as the most prolific architect of the century (Giddings and Horne 2002). However, it is not clear who was in this large office and who did what. Even with modernists such as Frank Lloyd Wright, Mies van der Rohe, Oscar Niemeyer and Le Corbusier, only a few other names from their offices occasionally appear in publications. This is in the context of many modernists expounding socialist principles. There was a change in the mid-20th century, when both practice and their architecture became more corporate. For the most part, this has continued well into the 21st century, but may be challenged by the Fourth Industrial Revolution. At the same time, globalisation has gained pace and led to the concept of starchitects. Their ubiquitous architecture is seen throughout the world. Several have offices in different time zones to enable 24-hour coverage across the globe. By necessity, they use digital techniques, as it is the only means of transferring design and technical solutions around the world.

Launched 42 years ago, the Pritzker Prize was initially intended to encourage greater awareness of how to perceive and interact with our surroundings. As society shifts towards responding to social, climatic and environmental emergencies, it appears that the Pritzker may be switching gear too. In 2020, the Irish architects Yvonne Farrell and Shelley McNamara won for their consistent service to humanity, the jury commenting that their buildings are good neighbours and environmentally responsible. Until now, ecological awareness in architecture has been recognised through certification, but usually not awards. So, working in this way has come with few accolades. Moreover, the most lauded of what is now known as starchitects are invariably men. Zaha Hadid was the first woman to win in 2004, and Anne Lacaton in 2021 was only the sixth. Nevertheless, this may also be changing, as 50% of laureates have been women since 2017; perhaps more significant may be to investigate the appropriateness of starchitects. There is a feeling of re-running 19th-century Imperialism, with an elite that has global dominance. There is a strong view that African architects of the future, for example, will need to fully understand its complicated history of their architecture, be sensitive towards societal needs and be innovative in creating relevant, practical and sustainable solutions, to provide uniquely African solutions (Designscape Architects 2023; Schensul and Heller 2011). According to Kililo (2015), the use of digital systems can actually be detrimental to creative minds. In his experience, architects and students produce conceptual designs more slowly, and it often leads to designers settling for what the systems can offer. It also re-opens the debate about whether digital systems are inclusive or exclusive. Lucrative commercial projects enable investment in the latest technology, provided the infrastructure is available. Socially orientated projects generate much smaller margins and limit that kind of investment. In addition, affluence varies greatly throughout the world, which in itself, has led to global dominance by the elite. Throughout history, society has been represented as technophilic, i.e. those who enjoy modern technology and technophobic, i.e. those who dislike modern technology. This has been a counter-productive categorisation, divisive and aimed at dominance and subjugation. It will also be unfortunate if creativity in design is diminished by systems aimed at building production, efficiency and profit. Perhaps the design and production of buildings will become more fragmented in future. Government requirements for the use of computer-aided design and building information modelling have been based on an almost stereotypical building project. Practice that does not fit into this mould may be significantly different. There may be a future that is not about integrated design teams but practice in different genres. For example, they may integrate design and film, or architecture and the gaming industry. Thus, digitised techniques may transcend traditional disciplines. There are already Master of Architecture programmes in Cinematic and Videogame Architecture. Students will develop skills and research agendas for future careers in architecture but also applicable in the film industry, videogame design and emerging fields such as the design of virtual reality and augmented reality environments and metaverse technologies.

The professions are a 19th-century construct and there is debate about their fitness for purpose in the 21st century. According to Susskind and Susskind (2015), in an internet society that is being extended by artificial intelligence, there is an ongoing transformation in the way expertise is experienced in society, which could eventually lead to a dismantling of the traditional professions. Dramatic differences are appearing in the employment patterns of people of working age around the world. Evidence is accumulating that young people are spending less time with each employer. Moreover, they are moving through different careers throughout their working lives. A differentiation may also appear between academic and professional education. While both are necessary, the former could be the province of universities, the latter might take place in practices, with modest attendance at universities. As professional education may become a minority activity, learning within practices could be viewed as a way for them to retain staff. In this context, digitalisation may not be as significant in academic education as it will be in professional education, especially as creativity may come to the fore in the academic arena. In addition, the pandemic has fundamentally changed work patterns. Working from home has seemingly had little impact on productivity, and so many practices have since moved to hybrid working. One approach is that each team works in the office for two days a week. Nonetheless, there is concern about longer-term impacts of home working, such as mental health. It is also asserted that more innovative design solutions evolve from teams working together in person. In addition, graduates and apprentices starting their careers will miss the kind of co-operative learning that has been one of the pillars of studio education. All these developments indicate a need and urgency for the architectural profession to recognise changes. If it does not become proactive rather than reactive, it could be under serious threat.

The Architectural Project

A project can be defined as a piece of planned work and an activity that is finished over a period of time and intended to achieve a particular purpose (Association of Project Management 2023). Thus, organisational and time management and meeting a goal are key characteristics, and it is collaborative enterprise that is carefully planned to achieve a particular aim. This involves human factors, such as teamwork. The term project is extensively used by numerous disciplines, and it can be perceived differently. Some disciplines recognise that it is about new ideas which are forward looking, often involving research, original and novel responses, and manifested through spatial configuration. In the built environment, it is about management, goal orientation and teamwork, in a social and political context. In architecture, there is also the concept of the prototype, as each architectural project has a unique set of circumstances. However, the promotion of individual buildings may not always be the most appropriate approach. It is arguable that locations should take precedence. It is the combinations of buildings and spaces that create places that are recognised by people and generate identity. Each new building should make a contribution to the place rather than just being considered as an individual project. Thus, the concept of a project for a specific purpose, with a completion date may not be particularly helpful concept. Moreover, in the Fourth Industrial Revolution, it is evident that flexibility is crucial. Uses will change at a much faster rate than in the past. It will no longer be feasible to demolish obsolete buildings and re-use will be essential. Places are really in process rather than a product. It is in these circumstances that city models were developed. Initially, they were physical models, often constructed in timber. Authorities, who were concerned about the coherence of their cities, inserted proposals into these models to make an assessment of their implications. At the beginning of the 21st century, there was unease about the cost of this process. As digital techniques became more sophisticated, physical models gave way to virtual city models. These can be extremely detailed and produce realistic effects. As proposals are increasingly developed with CAD and BIM, they are readily integrated into the model. Different iterations can also be produced more easily. Nevertheless, the more detailed the virtual model, the more expensive it is to develop, and a good level of detail is required to make them effective. Other considerations are the funding for towns and rural locations, in addition to cities, and whether it is a First World technique. It may questionable if it could be afforded in other parts of the world. There is also a concern about the need to import the technology into some countries.

The environmental control of buildings is critical in the contemporary debate on an architecture that contributes to the resolution of the climate crisis. This kind of performance, alongside other structural, material, organisational aspects of design, additionally to the knowledge generation on all of them separately, increases the complexity of design and construction. Collaborative decision-making is greatly beneficial. BIM is a powerful information tool that offers humans access to shared knowledge, which enhances the exchange of ideas and subsequent decision-making. It also opens up channels for non-human activity in a form of artificial intelligence. While this is very efficient, there is disquiet about diminishing human engagement with the process. It is interesting that recent international research programmes have noted that narratives often do not take into account traditional and indigenous knowledge and skills when defining sustainable solutions. These are difficult to identify if there is a lack of engagement with and recognition of local communities. Yet, local solutions may have a better chance of being grounded because they are close to the originators and implementers. By understanding the local origins of sustainable solutions, obstacles to achieving sustainability can be highlighted, challenged and overcome (Belmont Forum 2023). This re-opens deliberations about the definition of clients, and whether they might be users and the community as well the fee payer. It is difficult to adopt a moral stance when income is needed and therefore, it is usual that access to funding sources determines both the process and product. They tend to favour digital techniques for their efficiency in production. Hochscheid and Halin (2018) emphasised that BIM is not extensively implemented in some countries, and progress needs to be less on the technical and more on its why is this highlighted? organisational–managerial aspects. Significantly, they do not mention architectural design or the creative process. BIM is an overall process of generating a three-dimensional database in the form of a model of information that pertains to the design of a building. It overlaps with CAD and includes similar techniques during the design phase. The use of CAD can offer notable opportunities for creativity, and some of the most exciting forms and spaces from the late 20th century have been as a result of its implementation. Indeed, some of these designs may not have been even achievable with traditional drawing techniques. Although, it should be noted that they are generally individual building projects and not related to enhancement of a place. Alternatively, BIM is essentially a building production tool. In addition, there are ongoing ownership issues. The principle is that all the parties to the process have equal access to the three-dimensional model, and all make contributions to it. Yet, for reasons that range from intellectual property to professional indemnity insurance, this has not become a reality. Thus, ownership of the BIM model can dictate the priorities.

While CAD is adaptable, BIM relies more on established practices because they are systematised. As previously identified, there are already changes occurring in education and practice, and by inference, the architectural project. Analogue methods are inherently flexible. There have been calls for many years for a change of paradigm in the way that future practitioners are perceived as being educated in silos. This is part of the notion that design and construct were separated in the 19th century and need bringing back together. The assertion is that if education does not shift into a live brief, and life-long learning processes, BIM will have more drawbacks than benefits to education (Dobellis 2013). Nonetheless, education is deeper than merely mimicking practice, especially as it is difficult to predict the future of both. Perhaps one constant is that education is more about creativity and design quality than building production. Some tutors promote CAD rather than BIM due to concern over the library of components limiting creativity. The counterargument is that BIM prepares graduates for mainstream practice. In addition, it engenders experience in teamwork as architecture students can interact with other embryonic members of the design team. CAD or BIM and analogue are seen as being mutually exclusive, and that analogue represents the past, whereas BIM is the future. This may not be the most progressive narrative. There are different aspects of the process that could benefit from either. For instance, virtual city models aside, analogue is effective in establishing the characteristics of context and place, whereas CAD is particularly appropriate for exploring form and space. It is also a considerably easier way of testing designs and adopting an iterative process. This approach could lead onto the notion of integration in analogue and digital techniques. The relational proportions can be flexible, as it has been established that there is not a stereotypical project nor uniform locations nor equal access to technology.

Analogue and Digital Processes

The outcomes of a research project in Newcastle upon Tyne, the United Kingdom, are extremely informative about the relationship between analogue and digital processes. It investigates the notion that designers of the past have used analogue techniques, whereas those of the present and future may exclusively use computer-aided design. The data come from senior architecture students, rather than practitioners. Thus, the focus is on the quality of design and not building production. The essential thesis is that neither the exclusive use of analogue nor digital technology generates the most innovative design solutions, but interaction between the two methods can offer the effective designs, if employed in the most appropriate proportions at the different stages of the design process.

For over 30 years, architecture students have been increasingly using digital equipment. Yet, this is against a complex background that is not necessarily based on increasing the efficiency of their creativity and design processes (Casakin and Wodehouse 2021). The Fourth Industrial Revolution has engendered a pervasive use of digital technology. Young people are at the forefront, with a range of activities from social media to gaming. There has also been pressure from practice for students to use computer-aided design as means of obtaining employment. It can be seen as the first stage of introducing building information modelling into studio projects, which may morph into building information management, with the dangers of sacrificing design quality for building production. There has also been contradictory advice from tutors with some still supporting Lawson's (1998) statement that CAD is merely computer-aided drafting. This presents an inconsistent picture for use of analogue and digital techniques in architectural education, which is confusing for the students. It is, therefore, not surprising that the results of this study show that the students adopt erratic sequencing in their design activities that leads to unreliable development of their projects.

There is emphasis on the use of computer-aided design for visualisation rather than design development. The final presentations imply that the process has been CAD-led, but it disguises the reality that the designs have been mainly generated through analogue techniques (Tables 5.1 and 5.2). The study summarises the stages of the process as follows:

Table 5.1 Existing Methods.

Stage 1: Understanding	Very little use of CAD except for representation of the site
Stage 2: Definition	Almost exclusively analogue
Stage 3: Synthesis	50% CAD in relating to context and exploring design ideas, otherwise analogue
Stage 4: Analysis and Evaluation	Mainly analogue
Stage 5: Presentation	High proportion of CAD in relating proposals to society, incorporating technical resolution and representation

Table 5.2 Proposed Methods.

Stage 1: Understanding	High proportion of analogue, while CAD techniques increase as the data become more available
Stage 2: Definition	CAD use becomes more established as solutions adopt a more structured approach
Stage 3: Synthesis	CAD continues to increase to a majority situation, as spatial principles and exploration of design ideas involve iterative activities
Stage 4: Analysis and Evaluation	Iteration is seen as fundamental to integrating solutions, and CAD is the most efficient way of achieving it
Stage 5: Presentation	Envisioning takes place earlier in the process and can be restructured relatively easily through the CAD model. In the latter phases of the presentation, analogue is re-introduced to offer individual character to the proposals

The study highlights the advantages o to conceptualisation of the design by working with an intrinsically accurate digital model and establishes that students could benefit from applying it beyond visualisation. Therefore, the use of computer-aided design can integrate the previously separate activities of conceptualisation and visualisation into a single model. In doing so, it enables development and presentation at multiple scales. One of the complicated processes in analogue is the repeated need to move from the general to the particular and back again, in a re-iterative pattern, as it requires new drawings for each iteration at both scales. The model can be used at any scale and, as previously stated, has the flexibility to adjust to each iteration. In addition, the exploration of form, space and structure, amongst many other design layers can be accessed and experienced by students and brought together at any scale. Thus, the use of a digital model enables the process to blend the generation and interpretation of design ideas. Nevertheless, teaching requires new approaches that consider its effect on the design process. In relating to the modifications required in education, this study shows that there has been a lack of research and little formal guidance. There is only a limited literature that considers how analogue and digital might be used together, and the relationship of these methods to the design process (e.g. Dickinson 2011). As a result of this research, insights are offered into how they can be used in parallel at different stages and the positive effects on the design process.

Architectural Design Studio is a learning environment, where analogue skills play a significant role in the exploration of design solutions (Soliman 2017). Representation involves understanding and articulation, within which conceptual design and illustration are two inseparable acts for most students. Illustrating and model making enable iterative stages of a design solution in two-dimensional and three-dimensional forms. Particularly in the early design stages, as an analogue technique, illustration allows for rapid iteration and testing of design exploration. It can convert tacit knowledge into explicit knowledge, articulating design ideas. The use of physical models allows two-dimensional ideas to be formed and creates a three-dimensional visualisation of the design. The downside is that a new model needs to be made for each iteration. Digital methods require fixed geometrical points in order to create a digital model. Some people claim that it limits freedom of form. Nevertheless, all building designs should have an underlying geometry, and CAD and BIM demand a discipline that ensures it will occur. Moreover, digital models are endlessly flexible in that they can be readily adjusted with each iteration. In the process actually undertaken by the students, the timing is erratic and fragmented. Stage 1 was still occurring 75% through the total design period, with activities spread across several months. Partly because Stage 1 was taking so long to complete, and students were tending to jump to Stage 3 in an attempt to catch-up time. However, they soon realised that there were gaps in information and returned to Stage 2, some of which continued to almost 90% of the available time. Stage 4 commenced at almost halfway through the period and continued to the end of the programme. Stage 5 did not start until near the end of the process and focused on the final month. The proposed timing is considerably more structured, with Stage 1 complete by one-third of the time. Stage 2 overlaps with the previous stage and extends to two-thirds of the period. Stage 3 continues the progressive move through the months, and re-iteration is evident in Stages 2, 3 and 4. Stage 4 commences less than halfway, and Stage 5 starts at approximately two-thirds through the time and continues to the end of the total period.

This study shows that pressures to introduce digital techniques have not led to an organised process. In fact, the ad hoc approach left burgeoning professionals in a state of confusion. The primary reason seems to be the notion that analogue techniques belong to the past and have no place in the digital future. The outcomes demonstrate that an appropriate interaction of analogue and digital techniques may produce the most creative designs in an organised time frame (Morton 2023).

Conclusion

In architectural design, tools have always been a challenge. Balance or even conflict between arts and sciences may be evident, with perhaps the arts depicted by analogue techniques and sciences by digital methods. The conundrum of the need for each building design to be unique has been explored, together with the role of the architect in a digital world. In particular, it is considered whether individual architects are viable in this environment or whether teams are the only feasible way of working. Digital creative authorship may result in shared authorship, but there are issues about intellectual property and blurring professional roles in the 21st century project teams. It has been established that creativity can be taught and more common than usually assumed. The elevation of certain designers, including starchitects may therefore be for other reasons than just their creativity. They represent a value system based on monumentality, with a global consumerist power structure. Alternatively, socio-political priorities with the involvement of communities, is gaining ground.

Young practitioners are not only regularly changing their place of work, but there is accumulating evidence of intent to change careers throughout their working lives. They can see various activities where their creativity can be expressed, and view just being involved in architecture as a limitation on their aspirations. Digitisation is transforming the way expertise is being experienced in society. There is also a debate about the concept of the building project. There is increased emphasis on place and re-usability of buildings, and places are in process rather than being considered as products. CAD and BIM are greatly beneficial to the exploration of form and space, but analogue methods are more applicable to context and place. Digitisation can be used to enhance place through virtual city models but may not be applicable in different parts of the world, or in smaller locations due to the cost of generating them. There is also a role for traditional and indigenous knowledge and skills. There are four key conclusions. The first conclusion is that the emphasis should be architectural design quality and creativity, rather than building production. Secondly, there is a debate to be had about the future of professions, including the urgency for the architectural profession to appreciate the changes that are occurring to practice. Thirdly, it is about recognising the differences in the ways of practicing that can transcend traditional disciplines in countries and cultures in buildings and places; and most significantly, acknowledging that there is not a stereotypical building project. Finally, it is important to challenge the assumption that analogue and digital are mutually exclusive. It would be constructive position to accept plurality and employ each to its best advantage.

References

Association of Project Management. (2023). *What Is a Project?* https://www.apm.org.uk/resources/what-is-project-management/#:~:text=A%20project%20is%20a%20unique,an%20agreed%20timescale%20and%20budget.

Bartley, G. (2023). Creative being. *Philosophy Now*, 153, Editorial 3. https://philosophynow.org/issues/153/Creative_Being.

Belmont Forum. (2023). Collaborative Research Action, Climate and Cultural Heritage. https://www.belmontforum.org/wp-content/uploads/2023/04/CCH-2023-Call-for-Proposals-updated-30062023.pdf.

Boden, M.A. (1998). Creativity and artificial intelligence. *Artificial Intelligence,* 103, 347–356.

Brawne, M. (1992). *From Idea to Building*. London: Butterworth.

Carpo, M. (2011). *The Alphabet and the Algorithm*. Cambridge, MA: The MIT Press.

Casakin, H. and Wodehouse, A. (2021). A systematic review of design creativity in the architectural design studio. *Buildings*, 11(1), 31. https://doi.org/10.3390/buildings11010031.

Colomina, B. and Wigley, M. (2016). *Are We Human? Notes on an Archaeology of Design: Notes on an Archaeology of Design*. Zurich: Lars Müller Publishers.

DeLanda, M. (2013). *Intensive Science and Virtual Philosophy*. London: Bloomsbury Publishing.

Designscape Architects. (2023). *The Future of Architecture in Africa*. https://www.dscape.co.za/the-future-of-architecture-in-africa.

Dickinson, S. (2011). The appropriate balance between digital and analog. *Design Principles and Practices*, 5(4), 467–474. https://doi.org/10.18848/1833-1874/CGP/v05i04/38129.

Dobellis, R. (2013). *The Art of Thinking Clearly*. London: Sceptre.

Drummond, H. (2023). Creating cities. *Philosophy Now*, 153, 14–15. https://philosophynow.org/issues/153/Creating_Cities.

Giddings, B. and Horne, M. (2002). *Artists Impressions in Architectural Design*. Abingdon: Routledge.

Ginichiu, A. and Schoenmaeckers, A. (2018). Architecture and politics: the ideology of consensus verus the reality if dissensus. Special Issue Politics Too much or Not enough. *Studies in History and Theory of Architecture,* 6, 27–40. https://sita.uauim.ro/article/6-finichiu-schoenmaeckers-architecture-and-politics-the.

Hochscheid, E. and Halin, G. (2018). A model to approach BIM adoption process and possible BIM implementation failures. *Creative Construction Conference (CCC2018)*, Ljubljana, Slovenia.

Kéré, D.F. (2022). Media release announcing the 2022 laureate. *The Pritzker Architecture Prize*. https://www.pritzkerprize.com/sites/default/files/inline-files/2022%20Pritzker%20Prize%20Media%20Kit_1.pdf.

Kililo, M. (2015). The CAD enigma: the backlash within architecture in Africa. *Contemporary African Arabic Architecture*. https://www.archidatum.com/articles/the-cad-enigma-the-backlash-within-architecture-in-africa/.

Lawson, B. (1998). Towards a computer-aided architectural design process: a journey of several mirages. *Computers in Industry*, 35(1), 47–57.

Lockey, G. (2023). In praise of aphorisms. *Philosophy Now*, 153, 16–18. https://philosophynow.org/issues/153/In_Praise_of_Aphorisms.

Morton, D. (2023). *The effect of Computer Aided Design (CAD) on the design process of final year master's in architecture students.* Doctoral thesis, University of Northumbria.

Nicolai, B. (2013). New monumentalism in contemporary architecture. *Anglia,* 131(2–3), 297–313. https://core.ac.uk/download/pdf/212372083.pdf.

North, M. (2013). *Novelty: A History of the New.* Chicago: The University of Chicago Press.

Resnick, M. (2017). *4 Myths about Creativity.* George Lucas Educational Foundation. https://www.edutopia.org/article/4-myths-about-creativity/.

Schensul, D. and Heller, P. (2011). Legacies, change and transformation in the post-apartheid city: towards an urban sociological cartography. *International Journal of Urban and Regional Research,* 35(1), 78–109. https://doi.org/10.1111/j.1468-2427.2010.00980.x.

Sklair, L. (2012). Towards an understanding of architectural iconicity in global perspective. In G. Bracken (ed.) *Aspects of Urbanization in China: Shanghai, Hong Kong, Guangzhou.* Amsterdam: Amsterdam University Press. 27–45.

Soliman, A.M. (2017). Appropriate teaching and learning strategies for the architectural design process in pedagogic design studios. *Frontiers of Architectural Research,* 6(2), 204–217. https://doi.org/10.1016/j.foar.2017.03.002.

Stamp, G. (1982). *Great Perspectivists.* London: Trefoil Books.

Susskind, R. and Susskind, D. (2015). *The Future of the Professions: How Technology Will Transform the Work of Human Experts.* Oxford: University Press.

Vogiatzaki, M. (ed.) (2018). *Architectural Materialisms: Nonhuman Creativity.* Edinburgh: Edinburgh University Press.

Vogiatzaki, M. (2021). Lacaton and Vassal: how this year's Pritzker prize could spark an architectural revolution. *The Conversation.* https://theconversation.com/lacaton-and-vassal-how-this-years-pritzker-prize-could-spark-an-architectural-revolution-157636.

6

Inclusion, Equity and Diversity

Katya Veleva

Blush Cloud Diversity and Inclusion Coaching, London, UK

Introduction and Context

Before starting the discussion about inclusion, equity and diversity, especially in a global context, it is important to note that it can have very different meanings around the world. In the United Kingdom, for example, the following are carefully considered: Inclusion, Equity and Diversity is a single concept and therefore should be treated as singular rather than plural. Whether they have capital letters is debatable, but would not be incorrect. Not sure how many times it is stated in the chapter, but would be good to use captial letters if it is not too many times

- gender identity
- sex
- ethnic background and nationality
- faith and spiritual practices
- LGBTQIA+ community
- socio-economic background
- mental health
- the disabled community
- neurodivergence in people
- carrying responsibilities

Various parts of the world may have reached different stages in the conversation about inclusion, equity and diversity, so it is important to be mindful of this positioning and drive the conversation to be productive and expansive as opposed to focusing on negatives and feeding into blame and cancel cultures. It may be helpful to think about the journey of inclusion by referring to The Kubler-Ross Change Curve (Elizabeth Kubler-Ross Foundation 2023) while remembering that there is not always the same direction of travel. However, there is a desire to meet people where they are (see Figure 6.1). There is also a need to think about how to move the needle towards inclusion for all while being kind and meeting

Architectural Design and Management in the Digital Age: International Perspectives, First Edition.
Edited by Bob Giddings, Silvio Melhado, and Jenni Barrett.

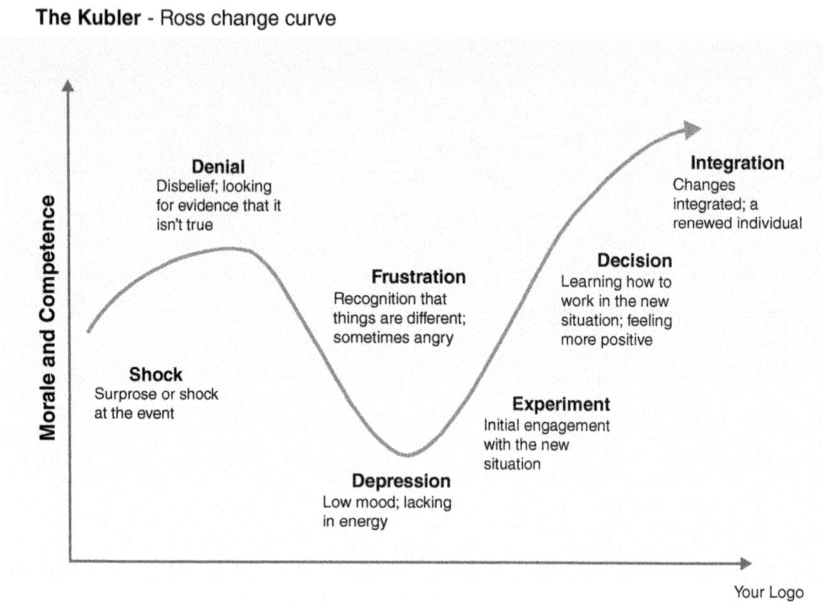

Figure 6.1 The Kubler-Ross change curve (with permission of the Elizabeth Kubler-Ross Foundation (2023)).

people where they are currently situated. Existing legislation must be understood, the power of the industry embraced and thought leaders challenge the law where it is insufficient or harmful. For example, when considering the construction industry in the Middle East, it is evident that people cannot even speak about diversity in terms of LGBTQIA+. In many of those countries, being queer is simply illegal, and in some places, it is even punishable by death. This is a gross violation of human rights and must change. Most important is that a conversation begins and that people are gathered together who are able to drive the changes to start. An alternative perspective is demonstrated in Australia, where it is noted that there is a positive trajectory overall, although starting from a low base (Galea and Jardine 2021). Another important aspect is that social justice, inclusion, environmental sustainability and economic progress are generally discussed as if they are separate notions. Yet on closer inspection, it can easily be seen that they are clearly interdependent, and improving each one of them in a thoughtful and long-term-oriented manner will support the others. For the construction industry, and specifically for architecture, there is an additional point. As Lennon-Johnson (2017) beautifully puts it, when she highlights that the built environment is inhabited by everybody for most of the time, people of all backgrounds must have a say in how the environment is designed.

Embracing and utilising technology in a balanced way can be one of the biggest levers for an industry to become truly inclusive and equitable. Omitting the need for physical space and attendance has the power to be a great equaliser. Working collaboratively in a digitalised way, can allow access to the profession to a greater number of people from more diverse backgrounds. There are, however, a couple of significant risks to inclusion, equity and diversity when digitalising. The simpler one is the fact that if marginalised people are already struggling to take up physical space in real-life workplaces and discussions, it may become even harder on a video call. Secondly, it can be more difficult to implement inclusion strategies when people are not physically together. So now, more than ever, it is necessary to reinforce inclusion practices. It is also important to remain vigilant about the possibility of isolation and overworking, especially in a profession and an industry that are already besieged with mental health and well-being issues. With the terrifying statistics around suicide rates on construction sites (Phoenix Health and Safety 2018; Thomas-Alexander 2021), it can be reasonably assumed that far more people are suffering from mental ill-health, but the toxicity of the culture prevents cases from being reported and treated. There is also the engine of perpetuation of grief, where a close colleague or even a friend has died from suicide.

Inclusion, Equity and Diversity in the Workplace of Architecture

Any workplace can benefit from an inclusive and diverse workforce. McKinsey & Company (2020) emphasises that consistently, year after year, the data show that gender-diverse organisations generally outperform others by 25%, and ethnically diverse companies are 36% more productive than monocultures. This happens because of many parameters that are improved by diverse perspectives. As the talent pipeline widens, staff gain energy by being able to be themselves at work, which improves retention. Representation motivates even more talent, and ultimately, the workforce has a better capacity for problem-solving. The talent shortage has concerned the construction industry as a whole for quite a while. Even architecture, despite its position as an attractive and prestigious profession, continues to see a rising number of professionals leaving due to its toxic culture and employee burnout (Hopkirk 2021). For any industry, more talent and boosting retention are vital. In conversations about employing talent, the cliché of always just hiring the best person for the job is often repeated. It is part of the discussion about creating a meritocracy. However, merit is significantly defined by privilege, so to look at it as an objective measure when considering inclusion, equity and diversity is a flawed concept. It might be more effective to think about hiring the best person for the team. This might offer perspective and experience that are lacking, and the challenge is how to shape the entry requirements to attract those people. When the arguments return to the ability to meet the demands of the work, it is worth considering those employees having to recover from adverse experiences in architecture, and re-examine those demands.

King (2015) has reported her experience as Chief Diversity Officer for British broadcaster Channel 4. She discovered with horror that they had not invested in any black film directors. She mandated her recruiters to find a black director in whom they could invest. After two months, the recruiters came back and said that there was not anyone who matched the requirements. When she replied that the search needed to continue, conversations started to escalate around a concern that standards were being lowered just to make space for a black director. Eventually, Steve McQueen was appointed, and Film4 backed 12 Years a Slave, which won three Academy Awards. King (2015) leaves others to ponder on whether standards had been lowered. Similarly, in architecture, there is a need to reconsider the requirements that have been implicitly set. There should be scrutiny of what they entail and how some of the inequities that grow from them can be mitigated (Waite 2020). There is anecdotal evidence that some of the most prestigious architectural practices in London are notorious for hiring only from certain schools of architecture. They claim that their focus is merit, so it is just a matter of who applies. It may be beneficial to evaluate the processes for entry and success in those few schools. There seems to be a contradiction in merit-providing education that is governed by ability to pay, and therefore merit as an equitable parameter for selection is open to question. The different cultures in schools of architecture are also worthy of investigation, as well as the coping strategies required by staff and students to survive them. The role of senior practitioners is also interesting, in terms of the kind of example that they wish to set and the influence they wish to exert. It may be valuable for managers in schools of architecture and senior practitioners to take a moment and write down their own reflections on culture and influence. The ability for students and staff in education and practice to be themselves significantly benefits productivity. Also, in situations that are unwelcoming, over 61% of employees put significant effort into changing who they are to fit into the workplace (Nottingham 2020).

As the conversation about social justice has evolved, so has the thinking and language. In recent years, the word equality has started to be phased out and replaced with the word equity because the focus has been refined from just the starting position to encompass the actual outcome. Equality means each individual or group of people is given the same resources or opportunities. Equity recognises that each person has different circumstances and allocates the exact resources and opportunities needed to reach an equal outcome. The comparison is visually represented in Figure 6.2.

Similarly, the word diversity is in common use, but it has been I learnt that the word may alienate the very people who have the most power and opportunity to enact change. In addition, research and practice have shown that diversity on its own does not offer significant improvements in the key performance indicators (Grace and Ammerman 2022). Thus, it is now used alongside inclusion and equity. The theory around unconscious bias reveals that the brain has evolved to view difference as a danger. This process of thinking and the

Figure 6.2 Comparison of equality and equity (Reproduced with permission of the Robert Wood Johnson Foundation, Princeton, N.J. / Lennon-Johnson, K. (2017)).

evidence for its verification have existed far longer in the Maasai on the Savanah, for example, than in the modern world with its relative safety and predictability. The rapid development of society within the long timescale of human existence has hard-wired the brain into a position where it is not meeting the needs and requirements of an inclusive collective. Instinctual and immediate behaviours generate feelings that are defensive, insecure and unsafe, when in the company of different kinds of people (Robbins 2019). These feelings are not helpful for creative and collaborative working; in fact, quite the opposite. Findings by Phillips et al. (2006) from a series of experiments to assess the capabilities of homogenous and diverse groups of people are an excellent illustration. The diverse groups categorically outperformed the homogeneous groups in terms of accuracy and quality of decision-making. The results showed that with more varied perspectives, a group has a more holistic understanding of a problem or a situation and will have better ideas and solutions. Yet, the diverse groups felt far less effective and confident in their work. This may be because of the hard wiring of the brain. What needs to be done, is to introduce a coordinated and intentional effort of not just accepting difference, but an active welcoming of difference. This will require training, fostering skills and productive discussion as part of lifelong learning.

The principal issue that has inhibited digitalisation within the construction industry is the complexity of contractual responsibility. However, there has also been the culture of the industry, which is a considerable barrier and not really contested. Due to the rigidity, strict vertical hierarchy and ultimately patriarchal power models, it has been problematic for a group of professionals from diverse disciplines to collaborate in real time. It requires significant emotional intelligence and is part of the overall issue of encouraging people from diverse identities to collaborate. The narrative around building information management is that professionals are already working in a cooperative way and that digitisation just involves more sophisticated technology. Yet, this is debatable. While the design team has always been fundamental to the process and the way it has been conducted, it is hierarchical and, some would argue, inefficient. In a traditional structure, it is based on the architect as team leader. Nevertheless, with BIM, there may be a need for a more egalitarian structure. There are parallels between this kind of inclusion and healthy collaborative work. It may be helpful to consider BIM as agile working adapted to the design process (Sakikhales and Stravoradis 2017). This is where people, processes, connectivity, and technology, time and place, are brought together to find the most appropriate and effective way of working to carry out a particular task. The overlap between digitisation and inclusive working is a constant experimentation-evaluation-iteration process. The more complex construction and engineering become, the harder it will be to continue to work within separate disciplines. In a parallel case, Gawande (2009) refers to a newspaper article describing a new sophisticated plane that exploded on the first test flight because it was too much of a plane for one man to fly. He opens with the idea that collectively, a point has been reached where there is too much knowledge for one brain to handle. So, to avoid errors caused by simply failing to recall information at the right moment, prompts are needed and generally, working together is better. There is an argument that there has always been more knowledge than one person can deal with; it is simply that people have not been clearly aware of it. Moreover, knowledge from a collection of personal lived experiences is a huge resource but has never been particularly well shared. It is often referred to as the gap in organisational learning.

Inclusion, Equity and Diversity in the Practice of Architecture

The process of designing the built environment in which everybody lives, plays and works is a responsible one. Yet, it is not clear that junior architects can feel that responsibility if they spend months assembling door schedules. Perhaps, they start remembering it during a spell of toilet design amidst a trans rights debate, but they may never truly take on the weight, accountability and humility that accompany it, when they are involved in menial tasks. One of the advantages of BIM, for example, is that it automates these kinds of activities and frees up space for all involved to consider higher-level issues. The benefits of digitalisation in architectural design can therefore reach a broad range of professionals.

Accelerated by the pandemic, one of the most obvious changes due to the Fourth Industrial Revolution has been remote working. It benefits a wide range of people and has opened up possibilities for people with mobility issues to take part in the workplace. Others with parental and caring responsibilities can gain time back from commuting, and people with neurodivergence who may not gel with busy travel and noisy open-plan offices can now demonstrate more of their abilities. Financial and carbon savings can also be made by not travelling. Even since the pandemic, there is an established pattern in large UK organisations of not requiring employees to be on the premises for more than 40% of their contractual time, and some permit 100% home working. Since the start of the pandemic, managers have had a notion of never going anywhere for meetings again. There are many benefits from the ability to work at home, but the situation needs more careful examination. In numbers of meetings in person, only certain people contribute, and others may not say a word. Also, the quieter ones are often from various marginalised backgrounds. In a physical meeting, a manager with high emotional intelligence who notices colleagues trying to make a point but not being able to, can make space for them. This is much more difficult in a virtual meeting. Moreover, this is only one of the ways in which people can start feeling extremely isolated by working from home. With junior staff, the learning experience can become slower and less stimulating if they are unable to ask questions and obtain the immediate support that they would like. So clearly, there is no equality for everyone working from home, but an equitable choice of work environment can be worked out if it is thoughtful and balanced. It is key to start thinking about inclusion in an additive mindset, not one of scarcity. It has become habitual to seek absolute answers and rules that simply apply blanket coverage. Conversely, the fundamentals of inclusion are built on adaptability and creating more options as opposed to making one size fit all. The type and amount of remote working may be the result of discussions between employees and their line managers. This is also an excellent opportunity to exercise, Transformational Leadership where leaders listen, guide, and motivate others to use change as a positive force to explore innovation and inspire action. This kind of leadership is a move away away from Transactional Leadership, where a group of people with power make decisions for everyone with no input from employees. This is outdated and has proven to be inefficient, but it is still evident in organisations. It is not possible to achieve an equitable, inclusive work environment if there is no understanding of the needs of employees, and that awareness cannot be developed without asking them. A transformational leader is in constant dialogue with employees and makes an active effort to comprehend their needs because different people simply require different help to thrive. Digitalisation may offer an opportunity for support, but clearly, a personal relationship is necessary. If a decision affects people's lives, it is best to discuss it with them. In her story, Steinem (2015: 177) interprets this situation as asking the turtle.

> *I took a course in geology because I thought it was the easiest way of fulfilling a science requirement. One day the professor took us out into the Connecticut River Valley to show us the 'meander curves' of an age-old river.*
>
> *I was paying no attention because I had walked up a dirt path and found a big turtle, a giant mud turtle about two feet across, on the muddy embankment of an asphalt road. I was sure it was going to crawl onto the road and be crushed by a car.*
>
> *So with a lot of difficulty, I picked up the huge snapping turtle and slowly carried it down the road to the river.*
>
> *Just as I had slipped it into the water and was watching it swim away, my geology professor came up behind me.*
>
> *'You know,' he said quietly, 'that turtle has probably spent a month crawling up the dirt path to lay its eggs in the mud on the side of the road—you have just put it back in the river'.*
>
> *I felt terrible. I couldn't believe what I had done, but it was too late.*
>
> *I took me many more years of organizing to realize that this parable had taught me the first rule of organizing.*
>
> *Always ask the turtle.*

The advantage of digitalisation is that it provides more options as to where work can take place. Movement to physical places takes more effort and more resources to accommodate increasing numbers of people, and each move comes with a price tag. In a digital environment, customisation can be more accessible. Of course, there are still challenges of adoption and understanding of certain pieces of software, but realising their potential can allow the time and resources required for effective and inclusive use, provided they actually work. The only role for the leader in this instance is to ensure that an environment is created in which people feel free to express their needs and have the opportunity to do so. While there is a perception of piling on more responsibility to achieve the status of excellent contemporary inclusive leader, there are some protective boundaries as well. Continuing to ask about people's needs, and ensuring that the space is welcoming to this kind of dialogue are very important. Inevitably, there will be employees who are not fully aware of their own needs or are not ready to face them. Some workplaces offer additional support in the form of coaching or even therapy to help employees raise their self-awareness, and this is very positive. Nevertheless, if leaders are not trained as coaches, or it is not part of their job description, they should not be coerced into taking on this activity. There are other situations where employees need professional support with mental illness, grief, addiction and so on, but these are not the leader's responsibility either. The leader needs to be aware of what the organisation can provide as support, but they should not take on challenges they are not equipped to resolve. The wisdom of Steinem's (2015: 177) *ask the turtle* is not just that people may know best about their own needs, but also that people's awareness is part of collective leadership. Even if somebody thinks that they can make a contribution to someone else's position and what they should do, if that person has not reached the place in their mind where they are ready to take on this help or accept that they have a certain need, it would simply be in vain.

Digitisation can be challenging for languages, particularly those that are under-represented on the internet and risk digital extinction. There is a danger that digital systems based on the three dominant world languages – English, Chinese and Spanish – may exclude other languages and their cultures (European Union 2021). There may be bias in these systems towards international style architecture, particularly where post-colonial nations are grappling with their own architectural language. Tannen (1994) highlights the relationship between culture and gender differences and the role that language plays in it. She asserts that for men, work is a competitive place and vocabulary builds status. For women, it is about a network of connections, and the vocabulary seeks to offer support. According to Rozenburg (Scarton 2020), the development of digital design in architecture is at a critical time. The potential for increased reliance on intelligent machines means that they need to be generated without insidious bias. However, there is little evidence that it is occurring. Despite creating a tool in 2018 to calculate possible gender bias in English-language architectural design terms, the profession has not responded. Rwelamila (2023) stresses the importance of ethnicity by assimilating the colonial architecture in Pretoria-Tshwane while moving forward with development that reflects diversity and heritage. This will necessitate the generation of a national architectural language, which will require huge effort and considerable focus, not least in its digital interpretation. Alternatively, Mbhokodo (2012) is convinced that the aim of the transformation agenda for the city is to attract world-class investments, which will likely produce a western style of international architecture through systems such as BIM. While there is a noticeable literature about social equity in the design of the built environment, there is little about social justice and fairness in access to the process of designing buildings. It is intimated in this chapter that architecture remains an elitist profession, in which commissions are often obtained through social status. It is unlikely that digitisation will change this situation. In that respect, it is neutral in terms of social equity for architects. Less equal may be the ability to purchase digital systems and their regular updating. This will become increasingly evident as commissioning organisations progressively demand the use of tools such as BIM in the design process. The size of practice is also an issue, with larger practices having more capability to afford this kind of technology. In addition, some parts of the world are subject to intermittent electricity supply and weak Wi-Fi access. Why does wifi need capitals and a hyphon?

The Way Forward

There is much that can be done to increase inclusion in the architectural profession. Indeed, there are so many aspects that it can become overwhelming. Nevertheless, it is important to make a start. One approach could be to deal with it in a similar way to a complex building design. Starting with the conceptual, which establishes a strategy and design, the following practical actions can be broken down into stages. As with a building design, each situation is unique, bespoke and different.

The Conceptual

To create positive change in any sector and in any aspect, considerable internal work within an organisation is required. There is an established literature about it (e.g. Burke et al. 2012; Mansaray 2019), and in the most synthesised approaches to the topic, there are a number of matters to consider. First, it is necessary to develop a high level of self-awareness. It should be remembered that it is not a one-person job, and help can be requested. Knowledge of the tools used is beneficial. In this case, the privilege, power and influence of technology have become contentious words in today's culture. Yet, they are neutral. It is what is done with them that defines their positive or negative impact on the world. To make an informed choice, there must be an understanding of the digital tools that can be used, their impact on different people and the resulting ability of individuals to perform, express themselves and be creative. Just as there used to be a roll of tracing paper available for students in a school of architecture,it was not so much for its transparency, but for the feeling of it not being so precious and for the ease with which it could be ripped-up, scrunched-up and thrown away. The present-day equivalent are digital tools that allow fluidity, flexibility and adaptability. They are efficient and lean, so often the way they are structured will prioritise those characteristics, at the expense of inclusiveness. A good characteristic of software developers is that they move quickly, so feedback on issues of this sort will receive timely support. Most of the large software companies consistently organise outreach activities. They understand that architectural practitioners will know the issues, so they can respond and enable the professionals to guide the process.

Understanding Artificial Intelligence will also become very important. Currently, it cannot be used to create new situations because it only works with existing data. When this kind of technology is used to improve process and speed, there must be awareness of the shortcomings that are already built into the data that are being fed into the machine. For example, if AI helps to inform a design, it relies on a representative selection of existing buildings, and the proposed outcome will almost certainly not be an innovative design. The future of AI is unknown. If it is permitted to develop freely, its learning capability will create more capacity than just re-ordering existing data. This may lead to original designs. On the other hand, there is considerable debate about where AI is leading and disquiet about developing machine thinking that may challenge human decisions. Similarly, if a mechanistic approach is used to help with recruitment and personnel selection, the existing data will probably be exclusionary. Even if there is a belief that an architectural practice is well directed and has excellent creative prerequisites for inclusion, the existing dominant overall culture of the profession could have an adverse influence on it. It is not clear at present whether AI, with the ability to learn, will reinforce prejudices or initiate a new direction. Alternatively, international protocols may prevent the kind of AI that is able to act independently, due to the risks of the outcomes and lack of accountability.

The Practical

Once the mindset is established, there are some actionable strategies that can be pursued. Staff can be asked to support inclusion efforts if there is something real and identifiable for other employees, as well as improving their own situation. African-American women, for example, only comprise 0.3% of architects in the United States (Pogrebin 2016). There is evidence of where only black women in offices are suffering from burnout because they are always asked to step into any diversity-related initiatives, along with their already full-time schedules as architects (Jackson 2022). Inclusion initiatives are real work, and they require significant skills – high-level communication, training, facilitation and mediation.

In the world of work, any marginalised people who speak out about the oppression they are experiencing are expected not only to be able to advise those around them and deal with the emotional responses from people belonging to the dominant culture, but also to be able to identify the systemic origins of the issues and propose and implement solutions. At the same time, these people may also be engaged in demanding roles on major architectural projects. This is the equivalent to having two full-time jobs. It is most likely that they will not wish to challenge the dominant culture and just leave the organisation. An alternative could be to introduce employee resource groups, but there has to be appropriate training, support and time to do that work and not an additional role.

Another action should be to determine who is missing from the team. As decisions impact diverse people, they need to be represented in the decision-making. If it is a large company, it should be a lesser problem. If it is a smaller organisation, it is not necessary to start employing people who are unaffordable, or worse, losing people, just to diversify. Nevertheless, it is essential to remain mindful of what is missing, whose perspectives are relevant and how these aspects should be considered next time someone new is invited to join. It is often the case that the leadership layer of an organisation is quite homogenous, and those involved in succession planning look the same. Again, it is not necessary to remove people's hard work; it is about seeking to add thoughts and views. Creating a shadow board of people from diverse backgrounds or making space for employee resource groups to input into decisions is an easy way to improve the inclusivity of leadership (Nishii and Leroy 2022). In the practice of architecture, this can be more difficult. In a creative process, the existing talent should not be judged on who they are Nonetheless, remaining mindful about the composition of new employees, where they come from and their experiences can be part of recruitment. Architecture is an elitist profession by Its nature. The length and cost of architectural education are two of the reasons why it tends to remain homogenous in terms of socio-economic background. Investing in apprenticeships and continuing education for existing employees may attract people who are different. In summary, it is about who are the best people for the team, and how to attract them to the team are the principal issues.

Regular review of internal policies to ensure they are helpful to a variety of employees is good practice. There can be outreach to external support for this or allow employee resource groups to help with it. In particular, there should be reflection on whether policies address and take reasonable care of people, including the following: not sure what is happening below but the hyphens are necessary

- genders – cis women, non-binary people, trans people and cis men
- sex characteristics – female, male, intersex
- sexual identities – heterosexual, homosexual, bisexual, pansexual, queer, etc.
- ethnic or national backgrounds – the make-up of the employee community and the proposed building user community
- faiths – building user community despite the make-up of architectural staff
- socio-economic backgrounds – not everyone among the junior staff has familial support; some of them may have to financially support their wider families
- caring responsibilities – parents, adult caregivers
- neurodivergence – autism, dyslexia, dyspraxia, attention deficit hyperactivity disorder (ADHD), attention deficit disorder (ADD) and others

- disabilities – mobility, but also hearing and visual impairments, chronic illnesses and others
- mental illness – depression, anxiety and others
- demographics – younger and older colleagues; do you have a menopause policy?

(Powers 2023).

Resources for inclusion are sometimes available and accessible in organisations, but they may not be enough. Regular updating and revisions, especially in growth contexts and the general environment becomes more inclusive, should be integral to the resource strategy. Perhaps even more important is that they are periodically brought to the staff's attention so that they are reminded about them. This is not just casual information being disseminated around the workplace; it is a change in mindset that demands repetition of the message. The details could also be provided in different formats to allow for disabilities and different learning styles.

Digitisation encourages members of staff to establish personal user guides involving how much each person feels included in the team. Staff can be encouraged to write-up various personal characteristics that they may wish to share with others. Colleagues could introduce themselves, the identity that they would like to communicate to others, inform them of the way they conduct themselves, and the way they work. They can say when they work most and least effectively, including physical environment, organisational style, and time of the day. They might identify if they are open to random phone calls or whether a written version of any information might be required. It would be good to know which employees prefer summaries of information or detailed accounts, whether the preferred method of offering information is discussion, phone, video or other means, and the mode for receiving feedback. Staff could suggest favoured ways of collaboration and some proposals for social engagement. Finally, there is the notion of speaking out and celebrating. Communicating internally and externally about commitments to inclusion can be useful. There may be cultural celebrations and days of remembrance. Having a calendar to follow is helpful, but the company should ensure that the above actions have been incorporated before celebrating the various awareness months. This will enable the practice of team building with an emphasis on getting to know and embrace differences as opposed to finding similarities.

Conclusions

It has been established that inclusion, equity and diversity have different meanings around the globe. Also, various parts of the world may not be at the same stage. For instance, the Middle East is in a very challenging period for LGBTQIA+ community. In terms of architecture, people from all backgrounds should have a say in how it is designed. It is argued that utilising technology in a balanced way can lead to inclusivity and equitability. However, if people feel marginalised in physical space, they may feel isolated in virtual space. Inclusive and diverse workplaces that demonstrate equity outperform others, and the architectural profession in particular suffers from staff turnover due to toxic atmospheres. Power structures are inhibiting widespread adoption of digitalisation. These systems are predicated on professionals working together in a cooperative manner. Moreover, computer-based systems can dramatically reduce menial tasks, and young architects may be able to contribute more to the designs. Yet, digital design has its own challenges in a worldwide context. Systems based on the three dominant world languages English, Chinese and Spanish may exclude other languages and their cultures. There may be bias in these systems towards international style architecture, particularly where post-colonial nations are grappling with their own architectural language. Small practices and consistent access to electricity and wifi are further inhibitions. The way forward involves providing capacity for different ways of working and not just assuming that digitalisation will happen in the future. with all other methods should be consigned to history. Digital technology needs to be more accessible for architects in less privileged situations, who would like to use it. This requires simplicity, less expense and coping strategies for poor infrastructure. To survive, the culture of architectural practice should become more inclusive and less based on power structures. Diversity of architects will enhance practice and produce higher quality environments. Digitalisation can encourage inclusion, equity and diversity, but it is not a panacea. It is only as good as the structure of the organisations that use it and should not be assumed to have exclusive use in the architectural design process.

References

Burke, C., Stewart, K. and Macdonald, I. (2012). *Systems Leadership: Creating Positive Organisations*. Aldershot: Gower.

Elizabeth Kubler-Ross Foundation. (2023). *Kubler-Ross Change Curve*. https://www.ekrfoundation.org/5-stages-of-grief/change-curve/ (accessed 9 November 2023).

European Union. (2021). *Digital Cultural Diversity*. European Parliament. https://www.europarl.europa.eu/RegData/etudes/BRIE/2021/690570/EPRS_BRI(2021)690570_EN.pdf.

Galea, N. and Jardine, M. (2021). *Building an LGBTQ+ Inclusive Workplace: A Blueprint for Australia's Construction Industry*. Final Report. Sydney: University of New South Wales Human Rights Institute. https://media.business-humanrights.org/media/documents/LGBTQ_Inclusion_Final_Report_22_June_2021.pdf.

Gawande, A. (2009). *The Checklist Manifesto: How to Get Things Right*. New York: Metropolitan Books.

Grace, B. and Ammerman, C. (2022). How business schools are really doing with diversity. In *Diversity, Equity and Inclusion*. Cambridge, MA: Harvard Business Publishing. https://hbsp.harvard.edu/inspiring-minds/how-business-schools-are-really-doing-with-diversity.

Hopkirk, E. (2021). Number of architects wanting to leave their jobs goes up to nearly half. *Building*. https://www.building.co.uk/news/number-of-architects-wanting-to-leave-their-jobs-goes-up-to-nearly-half/5111918.article.

Jackson, A. (2022). One expert's message to black women burned out at work: workplace DEI is a company issue, not yours. *CNBC.com*. https://www.cnbc.com/2022/07/06/expert-dei-is-a-company-issue-not-responsibility-of-black-women.html (accessed 10 November 2023).

King. O. (2015). *The Creative Industries: Why Our Future Depends on Them*. Speech to Parliament as Shadow Broadcast Minister. https://www.oonaking.com/house-of-lords/hol-speeches/item/297-the-creative-industries-why-our-future-depends-on-them.html.

Lennon-Johnson, K. (2017). *How Does Your City Affect Your Health?* Speech presented at TedxBrum (16 November 2017). https://www.ted.com/talks/kathryn_lennon_johnson_how_does_your_city_affect_your_health.

McKinsey & Company. (2020). *Diversity Wins: How Inclusion Matters.* New York: McKinsey & Company.

Mansaray, H.E. (2019). The role of leadership in organisational change management: a literature review. *Journal of Human Resource Management,* 7(1), 18–31. https://doi.org/10.11648/j.jhrm.20190701.13.

Mbhokodo, I.M. (2012). Heritage, an asset for the development of the capital, in City of Tshwane. In N. Clarke and M. Kuipers (eds.) *Re-Centring Tshwane – Urban Heritage Strategies for a Resilient Capital.* Pretoria: Visual Books. 27–37. http://www.braaksma-roos.nl/wp-content/uploads/2017/11/Re-Centring-Tshwane-Urban-heritage-strategies-for-a-resilient-capital-Nicholas-Clarke-Marieke-Kuipers-Job-Roos.pdf.

Nishii, L.H. and Leroy, H. (2022). A multi-level framework of inclusive leadership in organizations. *Leadership in Organizations,* 47(4). https://doi.org/10.1177/10596011221111505.

Nottingham, A.B. (2020). *Covered Potential: The Perceived Impact of Workplace Covering on High Performers.* Doctor of Education Dissertation, University of Pennsylvania.

Phillips, K.W., Northcraft, G.B. and Neale, M.A. (2006). Surface-level diversity and decision-making in groups: when does deep-level similarity help? Group processes and intergroup relations. *Psychology and Counseling,* 9(4), 467–482. https://doi.org/10.1177/1368430206067557.

Phoenix Health and Safety. (2018). Construction worker suicide rates are three times higher than the national average. https://www.phoenixhsc.co.uk/blog/construction-worker-suicide-rates (accessed 21 December 2022).

Pogrebin, R. (2016). I am not the decorator: female architects speak out. *New York Times.* https://www.nytimes.com/2016/04/13/arts/design/female-architects-speak-out-on-sexism-unequal-pay-and-more.html.

Powers, J. (2023). 39 types of diversity in the workplace you need to know. Builtin.com. https://builtin.com/diversity-inclusion/types-of-diversity-in-the-workplace (accessed 10 November 2023).

Robbins, S. (2019). The power of inclusion and mindful engagement in the workplace. *Breaking Bias, The Coca-Cola Company Workshop*, Atlanta. https://www.coca-colacompany.com/media-center/steve-robbins-on-power-of-inclusion-in-the-workplace (accessed 10 November 2023).

Rwelamila, P.D. (2023). Pretoria-Tshwane, South Africa. In B. Giddings and R. Rogerson (eds.) *The Future of the City Centre: Global Perspectives.* Abingdon: Routledge. 83–99.

Sakikhales, M.H. and Stravoradis, S. (2017). Using agile project management and BIM for improved building performance. In M. Dastbaz, C. Gorse and A. Moncaster (eds.) *Building Information Modelling, Building Performance, Design and Smart Construction.* Cham: Springer. 65–78.

Scarton, E. (2020). Hannah Rozenburg on tackling gender bias in architecture. *Australian Design Review.* https://www.australiandesignreview.com/architecture/hannah-rozenberg-on-tackling-gender-bias-in-architecture/.

Steinem, G. (2015). *My Life on the Road.* New York: Random House.

Tannen, D. (1994). *Gender and Discourse.* New York: Oxford University Press.

Thomas-Alexander, T. (2021). Suicides among unskilled workers rose in 2020. *Construction News Health and Safety*. https://www.constructionnews.co.uk/health-and-safety/suicides-among-unskilled-workers-rose-in-2020-10-09-2021/

Waite, R. (2020). Racism in the profession is getting worse. *Architects Journal Online*. https://www.architectsjournal.co.uk/news/racism-in-the-profession-is-getting-worse-aj-survey-indicates

7

Not Being There: The Potential for Digital Architecture to Reach the Disadvantaged and the Disenfranchised

Regan Potangaroa

Massey University, Auckland, New Zealand

Introduction

For architectural digitalisation to reach the disadvantaged and disenfranchised, it means that digital tools and techniques should be used to design and construct buildings that meet the needs and aspirations of communities. In order to be successful, these communities need to have access to the same advanced technologies and resources as more advantaged communities, and their voices and perspectives should be considered in the design process (Costanza-Chock 2020). This approach could lead to more sustainable and equitable buildings, as well as greater social and economic opportunities for those who have been historically marginalised. However, not being there offers a different twist, and digital architecture is considered to be not only working within communities but also as distant and remote. Such a situation is becoming more of a norm for practitioners such as architects, engineers, planners and humanitarians. Increasingly, they are working remotely through digital architectural tools. The potential for digital architecture to connect marginalised communities with physical architecture will be critically evaluated, especially whether digital solutions actually create more inclusive and accessible built environments.

Architectural technologies refer to the various tools, techniques and software used in architecture to design, analyse and construct buildings and other physical structures. While these technologies are often used by architectural practices, they are not solely for architects. Other professionals in the construction industry, such as engineers and contractors, also use them to plan and execute construction projects. Additionally, community groups and not-for-profit organisations can use the tools to engage and empower disadvantaged and disenfranchised people in the design and development of their built environment. There needs to be further consideration as to how this might be effective if the tools are used remotely. Arguably, the development of digitalisation started with the digital camera at the beginning of the 21st century. It was a particularly helpful piece of equipment and viewed as quite radical. Following the conflict in Afghanistan, refugees flooded into Pakistan.

Architectural Design and Management in the Digital Age: International Perspectives, First Edition.
Edited by Bob Giddings, Silvio Melhado, and Jenni Barrett.

A significant project was to construct refugee camps for approximately 320,000 people. A digital camera could record around 20 photos or about 60 seconds of 162 × 112 video onto a 1.44 MB 3.5-inch-high-density floppy drive. At the time, it was state-of-the-art, but 25 years later, it seems primitive. Yet, before then, the most sophisticated tool was a film camera, with which the film was developed, well after the event that had been recorded. In addition, there was always caution about taking a camera in such circumstances. By contrast, the digital camera was an immediate success, allowing for recording on-site conditions and incorporating photographs into reports, which in this instance provided direct video feedback to the United Nations High Commissioner for Refugees in Geneva, showing data in the field for those who could not be there (Kaur and Grassilli 2019).

Moving on several years, drone technology was developed. There have been significant concerns about the eavesdropping nature of drones. Deployment by the International Federation of Red Cross and Red Crescent from July 2011to December 2013 for assessment of areas impacted by earthquake in Port-au-Prince, Haiti was a particular case (International Federation of Red Cross and Red Crescent Societies 2012). There was disquiet about whether local people would accept this technology, as it may be seen as an invasion of privacy. The purpose was to ascertain where to build flood protection walls for the wet season. Drones made it immensely easier than having to track up the ravines through piles of rubbish, debris and sewerage. These examples highlight the essential dilemmas of digital technology. They create more effective working solutions for disadvantaged and disenfranchised communities, but they may evoke a loss of dignity for these people and the notion that outsiders are doing it for them and to them.

Context

The experiences of digital technology underline potential barriers to the reach of digital architecture, which could be perceived as manifestations of fear. Those offering this technology may have a number of concerns. It might be that practitioners are inhibited by the notion that digital solutions will not work as intended or will not be adopted by the communities for which they are intended to help. It can make practitioners tentative about taking risks or trying new approaches. They may also be anxious about unintended consequences, such as exacerbating existing inequalities or creating new ones. Practitioners may also fear the potential negative impacts of their designs and may be hesitant to introduce them without more research or testing. There could be apprehension about a backlash from the communities for whom they are producing designs (Hadadji et al. 2023). Thus, digital solutions may not be well-received or meet their needs. Some practitioners could be uneasy that their role in the design process will be diminished with the introduction of digital tools or that their expertise will be devalued. This perception might make them resistant to change and prevent them from fully embracing digital solutions.

There may be other concerns exhibited by the receivers of the digital architecture. Change can be unsettling, especially for those who have been living the same way for a long time. Digital architecture represents a significant shift in the way design of the built environment is conceived. Some may fear that this change will be too drastic and will disrupt their way of life. Those who are already marginalised or disadvantaged may fear that digital architecture will only serve to further exclude them (Gangadharan 2021). They may worry that they will not have access to the necessary technology or resources to participate in the digital revolution. In using digital technologies, there is a concern that decisions will be made by algorithms and artificial intelligence, rather than by people (Kolata and Zierke 2021). This loss of control over decision-making can be rather alarming, especially for those who have already been disenfranchised and may feel powerless. Hence, while digital tools can offer interesting solutions in an efficient way, the applications are so remote from the disadvantaged and disenfranchised communities that they would greatly assist, with both the givers and the receivers displaying inhibitions about the use of these tools.

The issues raised in this chapter will be illustrated by real-world situations. A practitioner perspective is adopted with a research lens approach to each situation. This approach will ground the chapter and enable a discussion about the advantages and disadvantages of applying digital techniques in different contexts and the ethical dilemmas associated with using advanced remote technologies. At the core is the question of how digital technologies can best serve the needs of disadvantaged and disenfranchised communities. An argument will be presented that digitalisation should not be the province of the developed world, but rather an opportunity to enable local initiatives and build capacity from within. By working in collaboration with local partners and stakeholders, practitioners can use digital tools to create buildings and build environments that are not just technically and aesthetically impressive but also socially and culturally responsive. The potential for digital technologies to transform the architecture of settlements and community development in challenging contexts will be evaluated. By providing a critical reflection on the opportunities and limitations of these tools, it is hoped that a more nuanced and effective approach to using digital technologies in the service of social and environmental justice can be developed. The chapter will conclude with a discussion of the potential future applications of digital technologies in community development. This will include a consideration of how advanced technologies such as artificial intelligence and virtual reality may impact the design process and how these technologies might be used to promote more equitable and sustainable outcomes.

The Issues

Emergency Response

Drone mapping can be used to assess the impact of natural disasters, such as flooding. Residents or refugees can be marooned on islands formed by the flood waters, and this situation might last for months. Disaster drones are all-weather search and rescue types that fly through the rain to determine where and at what level the response should take place. Prioritisation based on the video feed from the drone is key. Secondly, the drones are able to drop off payloads, such as emergency medical supplies, of up to 2 kg to people on the islands. Drone mapping has a number of advantages. The first is that it enables quick and accurate data collection. The drones collect a large amount of data in a short amount of time and with high accuracy. This is important because of the emergency influx that happens, and hence time is of the essence. This was the motivation for the concept of drone today-map tomorrow workflow. It is also more cost-effective than traditional surveying methods, requiring fewer resources and producing rapid completion. Drones can access hard-to-reach areas, reducing the need for staff to enter difficult locations and avoids risking their safety (Daud 2022). This form of mapping provides real-time data that can be used to make immediate decisions and adjustments to emergency response efforts. It also enhances coordination among the different sectors of organisations, such as United Nations High Commissioner for Refugees (UNHCR) and their implementing partners and suppliers, by providing a shared view of the area.

Nevertheless, the use of drones in any context raises ethical considerations and challenges (Wang et al. 2021). Approval has to be gained from refugee relief and repatriation commissions to fly them for mapping purposes over existing camps and proposed new camp areas. This is essential for clearances from police, military and any secret police operating in the area. Privacy is also a concern, as drone mapping can capture detailed images and information about people's lives and living conditions. To address this matter, the mapping team should ensure that the images captured are only used for the purposes of camp planning and not shared publicly or used for other purposes, without the consent of the community. Normally, this is enough to give them confidence that no invasion of privacy could happen. Moreover, there is experience that launching a drone is news that quickly spreads through the area, and it is not long before an audience assembles. Thirdly, there is potential for harm or injury caused by drones through accidents. To address this issue, flight protocols are set up, such as only taking off away from people and hovering at not below a 10-metre safety height. A flight plan is established that reflects weather conditions and especially the wind, recalling flights at any hint of rain or excessive humidity, not flying in low visibility conditions and introducing a safety zone during take-offs and landings. There are automatic safety settings on the drone, e.g. automatic landing if the battery is low, and object avoidance, such as potential bird strikes. The pilot needs to be able to land the drone if one of its four motors fails, requiring pilot vigilance throughout the flight. Thus, the flight team carefully screens anyone from talking to the drone pilot during flights. Safety is clearly a team effort. Finally, there are concerns about drones being used for surveillance or other purposes that could violate the human rights of the community. To address this possibility, the team must be transparent about the purpose and scope of the drone mapping and work closely with community leaders to ensure that the project will be conducted in a way that respects the rights and dignity of the community. The ethical considerations for drone use, need to be taken seriously and addressed through careful planning, consultation and collaboration with relevant stakeholders.

In 2017, the Rohingya crisis in Myanmar resulted in a mass exodus of over 671,000 refugees into neighbouring Bangladesh, joining 213,000 Rohingya from earlier years. The rapid influx of refugees created an urgent need for the development of housing and infrastructure in the overcrowded refugee camps (UNHCR 2018). Drone mapping was principally used for camp planning (Wellington School of Architecture 2018). Unfortunately, there was minimal local capacity in the communities, but it was quite significant in the UNHCR disaster teams. Also, there was significant cost and maintenance of the drones and the expensive computing equipment. However, the map products were an instant success with aid agencies. The emergency response in Bangladesh could have had significant potential for collaboration and partnership. Overall, the use of drone technology in the Rohingya emergency response demonstrates the potential of digital technologies to support community development in environmentally, socially and economically challenged parts of the world. By providing highly accurate and detailed data, is it not digitalisation that is used elsewhere in the book? can help aid agencies and governments develop effective and targeted responses to crises and support the needs of disadvantaged and disenfranchised communities. Nevertheless, the staff and technologies are imported, beyond the finances and expertise of these communities, and this results in others making decisions about their lives.

Rebuilding

TC Winston hit Fiji on 20 February 2016 as a Category-5 storm with average wind speeds of 233 km/hour and wind gusts of up to 306 km/hour. It was the first Cat-5 cyclone to directly hit Fiji and one of the most powerful storms ever recorded in the Southern Hemisphere. A total of 540,000 Fijians, or roughly 62% why are these items highlighted? of the population, were affected; over 30,000 houses damaged or destroyed and there were 44 fatalities (Australian Aid 2017). Fiji's Post-Disaster Needs Assessment states that a total of 495 schools were damaged or destroyed, at an estimated loss of 76.6 million Fijian dollars. The Fijian Government set up an adopt-a-school programme in which an individual national government or NGO could directly fund the rebuilding of a specific school. It was an innovative approach that immediately drew in international donors. One of those was the New Zealand Red Cross, which with funding from the New Zealand Government, selected Vunikavikaloa Ayra School on local advice (Fiji Red Cross Society 2017). The school serves a large catchment area of cane cutters, who are ethnic Indian. As part of their mandate, the Red Cross Societies insisted that the school be rebuilt using the local community associated with the school. Due to the cause of the damage, it decided that all 495 schools were to be built with increased storm resistance. It required larger steel sections that are not available in the country and would have had to be imported into the country. This presented a problem for the local engagement strategy, as the larger sections demanded different skill sets to those with which local people are familiar.

It was suggested that the use of FrameCad might provide a solution, as it is a system of light-gauge steel sections that has its own design software (FrameCad n.d.). The advantage was that it used light-gauge steel in coils that could be fed into an existing machine. The lightness meant that no cranes were required on site, and no heavy steel sections needed to be ordered. The FrameCad machine, in a 20-foot container, was sent to site from New Zealand, and all the light-gauge steel sections were fabricated and erected by the local people with assistance from FrameCad staff. In terms of community engagement, this system allowed for greater involvement of local people in the construction process. The software was relatively easy to use and allowed for the creation of building designs that could be easily shared and modified, based on community feedback. It was almost inexplicable that no local organisation or government department took up the offer from FrameCad to purchase the machine for other work in country. The lack of technology and knowledge transfer has been a major source of conflict between developed and developing countries, and this was an opportunity for nationals to carry on the work themselves with future projects after the aid workers had left. Moreover, it placed the nationals back in the situation where major projects would require the renewed importation of large steel sections and the need for a heavier-capacity construction sector from overseas. In this case, FrameCad allowed for the rapid construction of the building and resulted in a resilient community forming around the school by giving the community something that specifically needed to be completed in a much shorter time than traditional construction methods. Overall, the use of FrameCad in the rebuilding of Vunikavikaloa Ayra School on Fiji demonstrated the potential for digital technologies to support construction in challenging environments.

New Build

Hierarchical Decomposition is a design methodology devised by Alexander (1964) that breaks down complex systems into smaller, more manageable subsystems and components. It involves creating a hierarchy of subsystems, each of which performs a specific function and can be designed and tested independently of the others. Its great advantage is first that it is useful for sense-making when a situation is emerging. Second, it relies on people making connections or links between what are seen as the key factors and thus is ideal for situations where data are incomplete. At the time, computers struggled with problem decomposition, but with the computer power currently available, the analysis only takes seconds. In Myanmar from 2017, new warehouses were required for more efficient provision of relief supplies in Yangon, Sittwe, Myitkyina, Mandalay and Lashio. The problem in other countries had been that the warehouse designs had not met this objective, and therefore considerable pressure was exerted to ensure that the designs in Myanmar would be successful (Potangaroa and Hubbard 2017). It was decided to use Hierarchical Decomposition, as it permitted relationships to be tested and verified separately. This process began by asking staff from each warehouse to list the key aspects of what the warehouse should be able to do. They were coordinated by the national director, but it was the staff who needed to resolve what their new warehouse design should achieve. It was done in Burmese and in their cultural framework. The international team was then given the key points and how they linked with each other. The output was first an identification of the key issues or themes as seen by the users and secondly clustering of the design components for the warehouse facilities through the software. The spatial layouts of the warehouses, together with the site and boundary constraints, workflow and basic architectural design concepts led to the designs, which were presented as rendered drawings. Where there was an existing warehouse, the software was used to understand how it was working spatially, and the results also fed into the design data. Finally, the drawings were turned into contract documents by multi-national consultants, to be priced and built. These consultants have offices in Myanmar, and they are committed to bringing critical local knowledge together with global insights for the benefit of communities. The analysis, spatial design and warehouse workflow for each site resulted in the design of five different warehouses. Each had unique problems to be resolved. The one below was for Yangon (see Figure 7.1).

The Myanmar projects demonstrate the potential for collaboration and partnership between international organisations and local communities in the development of innovative solutions to humanitarian challenges. Yet, almost equally important is the role of problem-solving, which involves the use of computational tools, techniques and algorithms to help address complex and multifaceted problems. It is a clear demonstration of the reach and potential of such digital architectural tools. By leveraging the power of computational

Figure 7.1 A photo of the existing Yangon warehouse on the left and a render of the proposed warehouse on the right (Potangaroa and Hubbard (2017) with permission of Resilience to Nature's Challenges).

problem-solving patterns, relationships and insights can be identified that may be difficult or even impossible to detect using traditional methods (Caetano et al. 2020). The projects highlight the importance of engaging with local stakeholders as well as with the method of engagement. It was undertaken through the existing management structure within the warehouses and therefore reinforced rather than duplicated the way of working. It also directly fed input from the teams, at each of the five locations, into the decision-making process in an almost seamless way. A major consideration was ensuring that the projects did not perpetuate any existing power imbalances or further disadvantage any members of the community (Matera 2020). Additionally, the project team had to ensure that they were working with appropriate and transparent procurement processes to avoid any issues related to corruption or exploitation. They also had to ensure that they were adhering to all relevant laws and regulations, particularly regarding environmental and social impact. Finally, the team had to consider the safety and wellbeing of all involved in the project, including workers and the local community. These sorts of concerns are common to the application of computational tools, and this should be a signpost for future applications.

Housing Marginalised Groups

Home is one of the universal experiences of the human condition. Cooper (1974) notes that violation of a home is perhaps one of man's most deep-seated and universal fears, highlighting the importance that is placed on the house as an extension of the residents themselves. However, the architectural language of housing is changing. Driven by growing populations and increasing unaffordability, the demand for medium- and high-density mass housing has pushed the purchase of a traditional family house beyond young families. Thus, a disconnection exists between architectural designers, prospective residents, and the market affordability that affects the ability of housing to meet the needs of inhabitants. This is especially the case for marginalised families (Gaganpreet 2017). If virtual reality is to be used as a design tool, it is an interesting experiment to investigate how marginalised families would perceive their existing homes through immersive virtual reality. The homes were 3D scanned, and the reactions of the families observed (Musson 2019).

A Refugee Family in a Three-Storey Housing Block

Their initial reaction upon entering their virtual reality house was shocked laughter. The mother noted that her house was messy, while the father noted the broken balustrade on the stairs and the front door handle. These were details that they had been living and engaging with daily, but only noticed in the 3D-scanned environment. This was unexpected and very interesting.

Other responses were:

- intense interest in the details around the family space, which seemed to be a key area of the home
- a sense of safety in the communal corridor
- a feeling that their house was small but safe
- that the Virtual House was messy
- that the Virtual House was considered as a separate dimension and not a reflection of the actual home.

A key design aspect was perceived as the staircase, and how to create a place that could be inhabited, in addition to providing access to the upper level. It therefore had potential to generate accommodation by optimising space that was already there. A closed rather than an open balustrading was proposed, as well as inclusion of display spaces and seating along it, so the children of the house and those that visit could occupy the stairs and be in contact with what was happening with the adults in the lounge and kitchen areas, while at the same time being separate from them. The display space and the storage underneath addressed the messy element of the Virtual House. An impression was gained that traditional design methods would have required a higher level of sensitivity to the family and their home context to appreciate the amount of detail that was being identified by the 3D-scanned approach.

A Māori Family in a Single-Storey Timber House

A family of six permanent occupants spanning three generations lives in the house. It consists of two teenagers and four adults. One adult is retired. The household has a strong family network and community engagement. There is a steady flow of family friends and a good Māori cultural connection. An initial aspect was the variation in time that participants spent in the Virtual House. Occupant three was in it for more than thirty minutes, whereas occupants five and six were immersed for less than ten. However, occupant five suggested this was because she was unwell at the time. While there is a connection between time and data quantity, the level of detail provided by observational commentary was much higher than expected. Therefore, as opposed to individual interviews, the debriefing was undertaken as a family unit as everyone was already talking about the Virtual House and what they had found. The residents have extensive personal knowledge of the spatial details of their homes. Thus, any difference in geometric and textural would be noticed. Occupant three quickly identified that the Virtual House lacked realism. However, he suggested that because the model was not quite right, recognisable parts became important to contextualising himself in the environment. Thus, many details of the house stood out to him. He also appeared to be investigative when experiencing the Virtual House. This was observed in acts such as crouching to examine objects and moving closer to different areas. Photorealistic scans should be the standard, but this experience showed that all is not lost if there are some areas missed in the scans. Yet, it does mean that participants have to work hard to understand the environment, and where the scanning is not complete, participants have to interpret the spaces.

The technology certainly did reach both families, and there is potential for it to tailor projects for families. Despite living in the houses, there were aspects of them that were suddenly new, despite the familiarity. Moreover, design nuances that would not have been identified by other traditional methods were identified by the virtual approach. The spatial experience was heightened and emphasised. This is what Milk (2015) terms the ultimate empathy machine. It offers a different and more revealing picture of the relationship between family and home. These examples highlight the potential of digital architectural technologies to support the design and construction of homes that meet the specific needs and aspirations of disadvantaged and disenfranchised communities. However, there are issues of time and money. Each home is specifically designed for particular families, and the digital technology is expensive. This emphasises the notion of digital inequalities (Helsper 2021).

Summary of the Issues

In all the technologies explored, it is obvious that communities need to be able to access them if they are to realise the positive outcomes. Although used for the benefit of disadvantaged and disenfranchised communities, they were all owned and used by external agencies. Thus, each deployment of digital technology was for a very specific case, and there was no consideration of how it may become a continuing part of an ongoing process. The lack of technology and knowledge transfer has already been noted as a major inhibition to the development of these communities. The drone technology, such as that used in Bangladesh, will undoubtedly remain inaccessible because of security issues (Chang et al. 2017). This is the epitome of distant and remote operations. The Fiji school rebuilding programme was well intentioned in terms of rebuilding to higher standards, but the implications of a different technology that could only be imported and the lack of local skills presented considerable practical difficulties, higher costs and exclusion of the community. FrameCad appeared to be a great solution that would resolve these matters, and there was a strong business case for the purchase of the machine. It was a hugely missed opportunity that neither the Fiji Public Works Department nor the private sector took up the option. It demonstrated a deficiency in strategic thinking. The machine would have enabled a continuing series of projects to be undertaken without disruption. Not purchasing the machine indicated that thinking was short-term and only related to the immediate problem to be solved. Hierarchical Decomposition as a design method could be made available directly, but the issue would be having the staff with the necessary background to do the analysis. A better option might be setting it up on the internet and providing access to the technology via a bureau. Arguably the most significant aspect is the cost of the equipment. A few systems are becoming available on iPad that may address some difficulties with access, but generally, it will remain an issue.

It is also important to explore the role of digital architecture in fostering community participation and empowerment and how these technologies can be used to facilitate greater ownership among local communities (Skare and Riberio Soriano 2021). The standout technologies from those investigated are FrameCad and Hierarchical Decomposition. While the virtual reality had impacts on the families, there were issues of scalability. Drone mapping was an outlier. It had immense impact and did provide a tool that could be used for communities, but it did not involve community participation. Digital technologies can have reach, but they will undoubtedly have differing levels and modes of community engagement and empowerment. Clearly, not all technologies are equally applicable.

Finally, it is valuable to note the broader social and political implications of digital architecture, including questions around ethics, privacy and control over data (Susser et al. 2019); and how these issues intersect with issues of power and representation in the design and construction of buildings and communities. The use of digital technology is often shaped and influenced by existing power structures and social norms. This can result in technologies that reinforce and perpetuate existing inequalities and power imbalances, rather than challenging or disrupting them. The engagement with Hierarchical Decomposition was an attempt to avoid repeating previous situations with the repurposing of the warehouse designs. It was also the case for the FrameCad project, which backed up and eventually resolved what appeared to be an unresolvable situation, in which a school serving an ethnic Indian community was rebuilt to higher than current design and construction standards at a time of material and skill shortage to a budget and a tight deadline using a semi-skilled labour force. Both these complex projects were addressed by the respective technologies. In addition, as the virtual reality was focused on individual families, it would not be expected to have encountered such issues. However, it was not so clear for the drone mapping technology, where it was used to squeeze displaced populations onto land that had less than agreed international standards. Thus, the social and political implications of technologies do vary and need to be carefully and deliberately monitored and assessed to ensure they remain within acceptable parameters (Yin 2014).

The Future

The four digital technologies were chosen because they are examples of possible future applications that can reach disadvantaged and disenfranchised communities. These future applications could be in tandem and include the following:

1) Remote Sensing: Drones are one example of such technology with others being multispectral cameras for ground health, thermal cameras for tracking sub-terrain water and light detection and ranging (LIDAR) for 3D scanning of larger areas. These would be drone-based and can be used in tandem. Drones are already involved in the delivery of medicines but could have a bigger role in delivery and disaster response, asset management and machine learning with artificial intelligence to determine cyclone house damage automatically.
2) AI-Powered Design Tools: Artificial intelligence could be used to rapidly analyse data from different sources and generate design solutions that respond to local conditions and needs, helping architects and communities to design more efficient, sustainable and equitable buildings and infrastructure. Hierarchical Decomposition is one example of how it might be applied.
3) Digital Fabrication: Advances in digital fabrication, such as 3D printing, could enable local communities to produce their own building components and products, reducing dependence on external suppliers and creating more sustainable and resilient local economies. FrameCad is an example of that kind of technology.

4) Virtual and Augmented Reality: These technologies could be used to create immersive experiences that enable communities to better understand and visualise design solutions and provide feedback on the design process in a more engaging and effective way. The Virtual House is an example of what this technology may look like.

Overall, future applications of digital technologies for the disadvantaged and the disenfranchised hold significant promise for creating more sustainable, equitable and effective design solutions. However, it is important to ensure that these technologies are used in a way that is ethical, equitable and sustainable and that they consider the local context and involve local communities in the design process (Harrington et al. 2019).

Conclusion

Digital technologies have revolutionised the way buildings are designed, built and inhabited in the modern world. They have opened new possibilities to address some of the most pressing current challenges, including environmental sustainability, social inequality and economic development (Rosario and Dias 2022). Digital technologies have the potential to support disadvantaged and disenfranchised communities in the same way. One of the primary advantages is that they allow practitioners such as architects, engineers, planners and humanitarians to work collaboratively, and even remotely, to reduce the need for on-site visits. This is particularly important in geographically isolated and poor regions, where access to resources and expertise is limited. However, while digital technologies offer many potential benefits, they also raise important ethical and practical considerations. To address these challenges, it is important to take a community-centred approach to their use. This means involving locally based expertise in the design process rather than imposing outside ideas, as they could be part of the reason for any disadvantaging and disenfranchising of the community. It requires recognition of the importance of local expertise and incorporating it into the design process. Nevertheless, this local knowledge needs to be checked and verified rather than simply taken as fact. It is a delicate balancing act. Where possible, efforts should be made to ensure that digital tools and technologies are accessible to all communities, regardless of their technological infrastructure.

The four issues demonstrate how digital technologies can be used to reach the disadvantaged and the disenfranchised. The use of drone technology in the Rohingya crisis, FrameCad in Fiji, Hierarchical Decomposition in Myanmar and the Virtual House highlight the benefits of computational problem-solving in addressing complex challenges. These technologies provide valuable data, support construction in challenging environments, foster collaboration between international organisations and local communities and help design homes that meet the needs of disadvantaged communities. The challenges that remain are how to overcome and address any fears that block such technologies, and the bravery to accept the challenge and move ahead, but most importantly, how to achieve a situation in which the designs are generated by the communities and not imposed from outside.

References

Alexander, C. (1964). *Notes on the Synthesis of Form*. Cambridge, MA: Harvard University Press.

Australian Aid. (2017). *Tropical Cyclone Winston Education Response Evaluation*. Australian Government Department of Foreign Affairs and Trade. https://www.dfat.gov.au/sites/default/files/tropical-cyclone-winston-education-response-evaluation.pdf.

Caetano, I., Santos, L. and Leitao, A. (2020). Computational design in architecture: defining parametric, generative, and algorithmic design. *Frontiers in Architectural Research,* 9(2), 287–300.

Chang, V., Chundury, P. and Chetty, M. (2017). Spiders in the sky: user perceptions of drones, privacy, and security. *CHI '17: Proceedings of the 2017 CHI Conference on Human Factors in Computing Systems*. 6765–6776. https://doi.org/10.1145/3025453.3025632.

Cooper, C. (1974). The house as symbol of self. In J. Lang, C. Burnett, W. Moleski, and D. Vachon (eds.) *Designing for Human Behaviour: Architecture and the Behavioral Sciences,* Stroudsburg, PA: Dowden, Hutchinson and Ross. 120–146.

Costanza-Chock, S. (2020). *Design Justice: Community-Led Practices to Build the Worlds We Need*. Cambridge, MA: The MIT Press.

Fiji Red Cross Society. (2017). Students at Vunikavikaloa Arya School, in Ra province. https://www.facebook.com/FijiRedCrossSociety/videos/students-at-vunikavikaloa-arya-school-in-ra-province-have-been-studying-in-tents/1589665994379983/?locale=fr_FR (accessed 20 February 2023).

FrameCad. (n.d.). The world's most advanced steel frame building system. https://www.framecad.com/?utm_source=google&utm_medium=cpc&utm_campaign=brand_search_NEWZEALAND_AUS&utm_content=framecad_brand_terms&gclid=Cj0KCQiAgOefBhDgARIsAMhqXA7NKpGochScauZcRi6X-wjwGKalIAfh2Wr4FRAM4eDIcGXM5bqwUfgaAhoQEALw_wcB (accessed 20 February 2023).

Daud, S.M.S.M. (2022). Applications of drone in disaster management: a scoping review. *Science & Justice,* 62(1), 30–42. https://doi.org/10.1016/j.scijus.2021.11.002.

Gaganpreet, S. (2017). *Bridging Barriers: Study of Refugee Integration in New Zealand Communities*. Master Thesis Victoria University of Wellington, New Zealand.

Gangadharan, S.P. (2021). Digital exclusion: a politics of refusal. In L. Bernholz, H. Landemore and R. Reich (eds.) *Digital Technology and Democratic Theory*. Chicago and London: The University of Chicago Press. 113–140.

Hadadji, N., Toulan, N. and Dorra, M. (2023). Impact of digital architecture: the impact of digital technology on ecological formations and its effect on determinants of identity and culture in architectural design. *Journal of Engineering Research*. In press. https://doi.org/10.1016/j.jer.2023.09.004.

Harrington, C., Erete, S. and Piper, A.M. (2019). Deconstructing community-based collaborative design: towards more equitable participatory design engagements. *Proceedings of the ACM on Human-Computer Interaction*. 3(CSCW), 1–25. https://doi.org/10.1145/3359318.

Helsper, E.J. (2021). *Disconnect: The Social Causes and Consequences of Digital Inequalities*. London: Sage Publications Ltd.

International Federation of Red Cross and Red Crescent Societies. (2012). *Haiti Recovery Operation Plan of Action 2011–2013.*

Kaur, R. and Grassilli, M. (2019). Towards a fifth cinema. *Third Text*, 33(1), 1–25. http://doi.org/10.1080/09528822.2018.1546452.

Kolata, J. and Zierke, P. (2021). 'The decline of architects: can a computer design fine architecture without human input? *Buildings*, 11(8), 338. https://doi.org/10.3390/buildings11080338.

Matera, S. (2020). *The Role of Power Dynamics for Community Engagement: A Case Study in Myanmar*. Sweden: Lund University. https://lup.lub.lu.se/luur/download?func=downloadFile&recordOId=9016018&fileOId=9016024.

Milk, C. (2015). *How Virtual Reality Can Create the Ultimate Empathy Machine*. Speech presented at Ted Vancover BC (February 2015). https://www.ted.com/talks/chris_milk_the_birth_of_virtual_reality_as_an_art_form.

Musson, M. (2019). *Virtually Home; An Innovative Methodology for Identifying the Housing Needs of Marginalised Groups*. Master Thesis Victoria University of Wellington, New Zealand.

Potangaroa, R. and Hubbard, J. (2017). *Myanmar Warehouse Scoping Mission: Advisory Services for the New Zealand Red Cross Using Techiques Tested from and Through the Challenge December 2017*. Wellington: New Zealand Government. https://resiliencechallenge.nz/outputs/myanmar-warehouse-scoping-mission-advisory-services-for-the-new-zealand-red-cross-using-techiques-tested-from-and-through-the-challenge-december-2017/.

Rosario, A.T. and Dias, J.C. (2022). Sustainability and the digital transition: a literature review. *Sustainability*, 14(7), 4072. https://doi.org/10.3390/su14074072.

Skare, M. and Riberio Soriano, D. (2021). How globalisation is changing digital technology adoption: an international perspective. *Journal of Innovation & Knowledge*, 6(4), 222–233. https://doi.org/10.1016/j.jik.2021.04.001.

Susser, D., Roessler, B. and Nissenbaum, H. (2019). Technology, autonomy, and manipulation. *Internet Policy Review*, 8(2), https://doi.org/10.14763/2019.2.1410.

UNHCR. (2018). *Rohingya Refugee Emergency at a Glance*. UNHCR Mapping Unit. https://unhcr.maps.arcgis.com/apps/Cascade/index.html?appid=5fdca0f47f1a4649800 2f39894fcd26f.

Wang, N., Christen, M. and Hunt, M. (2021). Ethical considerations associated with "humanitarian drones": a scoping literature review. *Sci Eng Ethics*, 27(4), 51. https://doi.org/10.1007/s11948-021-00327-4.

Wellington School of Architecture. (2018). *Drones Help Develop the World's Largest Refugee Camp*. Victoria University of Wellington. https://www.wgtn.ac.nz/architecture/about/news/news-archives/2018-news/1682438-drones-help-develop-the-worlds-largest-refugee-camp.

Yin, R. (2014). *Case Study Research Design and Methods* (Fifth Edition). Thousand Oaks, CA: Sage.

Part IV

Emerging Concepts

8

Standardisation, Components and the Bespoke: New Design Challenges

Aya Jazaierly[1], Monia Del Pinto[1], and Fang Zhou[2]

[1] *Loughborough University, UK*
[2] *University of Science and Technology, Wuhan, China*

Introduction

When approaching the topic of standardisation and the bespoke, the first matter that brings itself to the discussion is what kind of standardisation and customisation are being considered. They could be related to the process and/or the product. It may be necessary to differentiate between two types of standardisation and customisation, i.e. the process of designing, drawing, modelling, manufacturing and constructing a project, and the product pertaining to the building and its components. Both definitions overlap and feed into one another, but it is important to draw a line between them to better understand how standardisation and customisation influence the way buildings are conceived, designed and made.

This chapter focuses on Building Information Modelling and the role of the ISO 19650, which cover the entire life cycle of architectural projects. Within this framework, when focusing on the product, BIM software concentrates on creating an object with measurable targets, such as number of square metres, standardised building components, market demands and sellability; in accordance with pre-defined customer sectors. It creates and generates optimised solutions following pre-defined parameters to achieve these targets to create the product. Shifting the focus onto the process, questions around the role of BIM software become channelled towards understanding the series of activities undertaken to create a project. When looking through this lens in the context of Architectural Design and Management, there is the matter of standardising the way that design then relates to the organisational and managerial structures within the context of the Architectural, Engineering and Construction (AEC) industry. More specifically, it touches on the roles of project collaborators, the way they communicate, take decisions and create the project and its components. In a broader sense, the standardisation and customisation of the process and the product bring the issue of design value into the foreground. Instead of only looking

Architectural Design and Management in the Digital Age: International Perspectives, First Edition.
Edited by Bob Giddings, Silvio Melhado, and Jenni Barrett.

at measurable targets, the issue of design value encompasses urban, social, cultural and political contexts that affect the design thinking. This requires a holistic approach to tackling design problems based on links and interrelationships between complex and sometimes non-quantifiable issues affecting the built fabric. There is a debate as to whether it is possible to translate the intangible aspects of a project into digitally quantifiable parameters. If so, the impact on the architectural and cultural outcome would require examination. Also, the impact on the AEC industry as a whole and the role of the architect within it would need re-assessment.

This chapter seeks to evaluate these aspects through the presentation of standardisation and customisation, in reference to the established frameworks. It demonstrates contemporary applications and moves to reflect on their design value in reference to the UK Design Value Framework through a historical overview. It starts by introducing the concept of building standardisation during the modernist period. Next, contemporary approaches to standardising the process and bespoking the product are discussed, with an analysis of the application of ISO 19650. Finally, a reflection on the implications for design value is carried out, and links to emerging technologies and trends are presented in the conclusion.

Standardising the Product: Modernism

Standardisation as an Architectural Solution

The first implementation of standardisation in architecture, intended as standardisation of building components, can be traced back to the 1920s with the large-scale social housing projects built across Europe. Rooted in 19th-century philanthropic planning, social housing projects represented an architectural response to the needs of the working population in a new industrial society. They evolved into the existenzminimum, a theoretical manifesto to guide modern architects' reflection on the production of affordable living units for the working classes, which embodied the changed approach to dwelling design, production, distribution and use. Design research around standardisation was brought forward by exponents of the Modern Movement such as Walter Gropius, Friederick May, Otto Haesler and Bruno Taut in response to a specific social need: Specifically, the demand for affordable housing for low-wage earners was affected by the rise in rent prices following German hyperinflation in 1923. This research was aimed at rationalising design and production and informed by functionalist principles. This was a common departure point that was then interpreted from politicised perspectives by different exponents of modernism. After modernism, the use of minimum spaces, and prefabrication have been reinterpreted in a variety of contexts until the present day, from social housing to temporary dwellings in post-disaster areas (Johnson et al. 2006). Most importantly, the need for consensus evolved into the International Organisation for Standardisation (ISO) standards informing contemporary design at multiple scales.

Early 20th Century Modernism

The modern standardisation in architectural design originated from the socialist dwelling culture rising in Europe at the beginning of the 20th century. In this context, the optimisation of spaces to maximise buildable surfaces and the large-scale edification of social housing estates were architectural responses to the housing needs of the working-class population. The rationalisation of architectural design applied from the new urban layouts to the living units, generating new forms meant to serve and embody and eventually influence the industrial society (Hillier 2007). Within this perspective, the replicability of forms and components became a key aspect in both urban planning and construction. Living units were composed of modular components, and multiple units were aggregated to form large housing estates. The rationalisation of building design and its processes was made possible by the introduction of design standards and by the large-scale mass production of prefabricated building components. Product standardisation from building elements to living spaces, and the design of housing units were assisted by codified standards. Among the most relevant examples is the 1925 Ernst May's Frankfurt Siedlung. Product standardisation was applied to concrete panels used in walls and floors, to frames for doors and windows and to kitchen spaces with the Frankfurter kitchen designed by Margarete Shutte-Lihotzky. Prefabricated panels and frames had an impact on the building process with optimisation of costs, buildability, and maintenance. The Frankfurt kitchen was an example of impact on the everyday lives of users. By interpreting Taylor's industrial methods, the Frankfurt kitchen offers an articulated example of a standardised product designed to optimise and rationalise use of space. In line with the intentions to emancipate and liberate users through rationalisation of spaces, functions and time – the solution was meant to be implemented on a large scale in the suburban Siedlung settlements. By applying the logic and methods of industrial production in architecture, modernist architects played the two-fold role of designers and social reformers. They optimised the building process while, at the same time, interpreting the spatial needs of a changing society. They responded to demands for affordability and salubrity of working-class housing and the demand for efficiency in construction.

Standardised Modular Buildings: Implications on Design Value

A retrospective assessment of the design value of early product standardisation can be performed by adopting the sociocultural, financial, environmental and democratic indicators of design value developed within the UK Design Value Framework (Drew et al. 2022). A distinction is needed between the approach and its outcomes. For example, the conceptualisation of house as a self-contained productive unit is linked to women's emancipation in designers' intentions. It retains strong sociocultural and democratic values, initiating a change in perception of the domestic space that would inform design theory and future architecture. Similarly, the rationalisation of the design and construction process and the consequent impact on architectural language, design theory and fabrication retain strong financial, environmental and sociocultural values as forerunners of contemporary practice. The changes that standardisation produced in the design process and architecture in general can be appreciated if the notion that standards survived the failure of modernism is taken into account. In fact, even after modernist principles proved inadequate to solve societal problems through built form as initially proposed, standardisation remained key in architectural design and product design worldwide.

Early examples of the shift towards product standardisation can be found in the publication of manuals such as the Bauentwurfslehre by Ernst Neufert, published in its first edition in 1936 and translated to English in 1970 as Architect's Data. The manual was envisioned for educational purposes but over time became a reference for professionals as well and is still in use. The relevance that standardisation acquired was also shown by the growth of national and supranational institutions to regulate standards and pave the way for process standardisation. In fact, the acknowledged improvement in efficiency enabled by standards and the need for a unified approach led to the establishment, in 1946, of the it would be helpful to use the full title at intervals to remind the reader what it means and to give a little variety, ie not seeing ISO repeatedly ISO, which gathered delegates from 25 countries and with 67 technical committees. ISO was born from the union of the International Standards Authority (ISA) and the United Nations (UNSCC) with the intention of creating a new global international standardising body. The first ISO Recommendation was ISO/R 1:1951, today referred to as ISO 1:2002, published in 1951 and titled The Standard Reference Temperature for Industrial Length Measurement. It was followed over the next decade by publications in mechanics, plastics, technologies and energy, to move in the 1960s from methods and terminologies to performances and safety and health aspects of products. With more than 24,600 standards released, today ISO is a global network of national standard bodies that establishes consensus-based, market-relevant international standards that support innovation and provide solutions to global challenges (ISO 2023). Among the most recent standards for architecture is ISO 19650, based on the British Standards BS1192 and the Public Available Specifications PAS1192-2, aiming at managing information over the whole life cycle of a building using BIM similarly it would be good to see the full name at intervals for the same reasons as above. The ambition of this standard is to not only reduce construction costs but also to enable efficient information management within global projects, creating a shared language for designers and contractors worldwide.

ISO and Standards for Modern Society

As with other international organisations having a global character, ISO was established after the Second World War, embodying the spirit of creating a better world by overcoming differences and sharing values. The guiding principle of friendship among equals in the 1946 headline of the ISO declaration shares the same spirit as the 1948 Universal Declaration of Human Rights, promoting understanding, tolerance and friendship among all nations. Besides the need for alliances to preserve balance and peace, these are all born out of the need to establish order through classification and categorisation. Giddens defines the need to categorise as a side effect of modernity in response to the perceived uncertainty, particularly evident after the Second World War: The creation of institutional labels and processes creates the illusion of a restored order and control. In this new scheme, the role of experts became central, as expert knowledge would help manage the challenges of the new society in a manner that would be safe, manageable and foreseeable. The growing importance of expertise led to the gradual compartmentalisation of knowledge and further specialisation within each discipline, creating a sense of safety and risk management. At the same time, it also generated disconnect between the expert and the general, non-expert public. Within this system, non-experts have no agency due to their lack of expertise, with all decisions being elaborated and entrusted to experts who know. In this way, the establishment of ISO fits within the paradigm of expertise within modernity but not with the contemporary accessibility of modern technologies.

Standardising the Process and Bespoking the Product: Contemporary Approaches

Design Process Through Digital Design Workflows

With the emergence of digital technologies and their gradual integration into the design discourse, architectural practices needed to reshape their design processes to move from the act of individual or even collaborative design to accommodate more information-based activities centred on data exchange. Architectural practices consequently moved to adopting digital design workflows instead of producing two-dimensional drawings that involved architects, designers and engineers from different fields. Thus, instead of documentation of the design process, digital design workflows integrate data gathering, spatial simulation, documentation and coordination within the design process (Deutsch 2015). This shift towards digital workflows came as a natural evolution of advances in digital technology, since the introduction of Sketchpad in the 1960s and the early versions of Computer-aided Design (CAD) in the 1970s. The adoption and integration of a multitude of digital tools is now the norm within architectural offices and the overall AEC industry. Practices now organise the design and execution of projects around digital workflows, and personnel are hired for their proficiency in digital tools to fit project needs. Those such as Zaha Hadid Architects, Frank Gehry, UN Studio, SHOP Architects and others were pioneers in this field during the 1990s, as they were able to design, draw, model and execute projects with bespoke geometry using complex and collaborative workflows.

The ISO 19650 Standardising Process

Although BIM was introduced as early as the 1970s, its actual wider use did not take place until the early 2000s, at a time when the architectural design process became more reliant on data-driven components. This was when the architectural office became a fully information-based industry. Allowing for collaboration with multiple disciplines, BIM presented the opportunity to close the gap between architects and engineers on one hand and between design and construction on the other. It allowed projects to be spatially simulated, integrating design input of all consultants into one consolidated, shared, intelligent virtual model. The idea, while appealing to all the protagonists within the architecture and engineering professions, necessitated new information organisation and management systems. New questions came to the forefront, especially about the control of parts of the model and when and how it is updated. The storage and management of complex and large amounts of data were debated, along with the security of a centralised virtual model. It is necessary to ensure that all collaborators are compliant with the same design standards, as well as provide effective and successful collaboration between the different parties. The organisation of team members' tasks as part of the project workflow is important, together with the management of seamless interoperability. From this perspective, and in order to manage the emerging design and technical challenges, ISO 19650 was introduced as the standard that would provide specifications for the concepts and principles, delivery and operation, information exchange, security and health and safety within the BIM process. It addresses all stakeholders involved throughout the project life cycle, including project owners, tenants, building managers, architects, engineers, quantity surveyors, building contractors and sub-contractors, manufacturers, specialists, authority representatives, end users and others. ISO 19650 provides a specific definition of all these stakeholders and offers a defined framework for the interaction, coordination and decision-making between them at all stages of the project from design conception to operation (ISO 2023).

At its heart, ISO 19650 allows stakeholders to access, input and manage information throughout the project life cycle, based on their needs and assigned roles. This includes not only the project's three-dimensional model but also information related to materials, cost, schedules, protocols and notes. It provides a specific set of definitions for all stages, procedures and parties in any given project, which in ISO terms is referred to as an asset. These definitions create the linguistic framework within which the overall information management takes place, and users of ISO 19650 must therefore fully incorporate this framework within the structure of the project. This becomes the basis for all the data exchanges, permissions and control over the BIM landscape, and consequently defines the information-based product that all these stakeholders are seeking to create.

Implications on Architectural Profession and BIM Washing

Within the context of ever-evolving digital workflows and standardising protocols, new issues arise about a shift in the architect's role. It may be that future architects will require a different type of education. This will have implications for architectural education, practice and eventually the whole built environment. Architectural practices operate within a defined hierarchy to design and deliver projects. Fitting with this hierarchy and collaborative context, the BIM process and ISO 19650 allocate specific tasks, scope and level of control over the digital model at different stages of the design process. While the BIM process aims to minimise losses and eventually reduce design and construction costs, practicing architects argue that this dissection of the process causes disruptions within the architectural process. Respondents report that *the disruptive nature of BIM can lead to increased dissociation of senior staff from the methods of design conception, development and document production* (Holzer 2016: 76). It is also argued that the increasing protocols somewhat cancel the iterative nature of the design process and constrain all participants in predefined roles and microtasks. With the BIM process, new roles are emerging, such as *BIM content creators/librarians, model managers (project level), BIM managers (office/organisation level), BIM coordinators (multidisciplinary project level) and BIM modellers* (Holzer 2016: 76). The ISO standard dictates ample detail and rigour around each step, whereas the creative process is inherently organic and less precise. It is open-ended and unpredictable, where ideas come together to eventually create the outcome. The design process itself is a series of not-always-measurable aspects, comprising the essential relationships between these aspects, intuitively working together for the design to emerge. Within the existing patterns of design procedures, architects and other designers constantly move between different design tools, including manual drawing, model-making, digital drawing and digital modelling, to explore their ideas visually, spatially and aesthetically (Brawne 2003). This non-linear process innately edits out and creates hierarchies in reference to the project conditions favouring specific aspects, focusing on them to generate the outcome. The BIM process, however, accounts for the aggregation of pre-defined elements within prescribed procedures to create the outcome. This pertains to the modelling and set up of BIM objects, which construct the overall project BIM landscape used by the relevant collaborators, restraining the designer's freedom to edit out, select and test ideas without accuracy because every step is prescribed and rigid.

Customised Building Components

When considering bespoke solutions, reference is generally made to the customisation of project components to fit users' needs, environmental and ecological footprint, structural optimisation and specific aesthetic features. These needs translate into parameters that, according to an identified criterion, would allow for the generation of a series of variations, which would in turn translate into form. Thus, it provides a geometrical interpretation of established inputs to a specific design within a digital model. Parametric workflows consist of command sequences enabling exploration, control and rationalisation of the design process and its outcomes. Within parametric design, customisation is achieved by incorporating additional steps and software plug-ins into architecture and engineering workflows. They integrate parametric design tools to generate different variations according to a set of variables, allowing designers to customise building components towards the outcome, i.e. a project feature, specifically tailored for a location, user or structural system. The debate around parametric design and its application to buildings has been widely explored and discussed by architects, theorists and historians in the past 20 years. Schumacher (2008) claims that parametricism is a style that reflects the society's diversity and possibility to differentiate itself:

> *Architecture finds itself at the mid-point of an ongoing cycle of innovative adaptation – retooling the discipline and adapting the architectural and urban environment to the socio-economic era of post-Fordism. The mass society that was characterized by a single, nearly universal consumption standard has evolved into the heterogenous society of the multitude. (Schumacher 2008).*

Carpo (2016) argues that digital parametricism opens the possibility for architects to experiment with a multitude of options, breaking away from the monotonous standardisation brought forward by modernism and bringing back the imagined diversity and variety in classical and medieval times. Allen (2005) also believes in the possibilities of digital tools, where they provide defined forms and are able to adapt to emerging needs relevant to contemporary lives. The use of parametric design starting with starchitects such as Zaha Hadid and Frank Gehry has become increasingly accessible to architects and other designers in the AEC industry. It has been utilised for optimising structural solutions (Brensing and Schittich 2013), customising patterns on building facades (Kuma 2009), and has had applications in customising construction based on the users' preferences (Eid Mohamed and Carbone 2022; Piroozfar and Piller 2013).

Implications on Design Value

It is important, within this context, to reflect on the design values offered by both standardisation and customisation for the design process and product. The UK Design Value Framework proposes four interconnected value domains – sociocultural, democratic, planet/environmental and financial/economic (Drew et al. 2022). The framework is meant to capture both positive and negative impacts on these domains through a value map and a value assessment table. The map outlines domains and the mechanisms of design, producing impact through specific activities. The table provides an overview of the impact assessment through defined indicators. It is important to mention here that these four domains are intertwined, as they all overlap and influence one another.

Impact on Sociocultural Design Value

From a sociocultural perspective, the impact of standardisation of the design process could manifest itself and emerge in the context of new architectural practice. It has been shown previously through implications on the architectural profession and BIM washing, that a standardised process could contribute to shaping a new architectural culture, where the role of the client and the IT technician would gain agency at the expense of the role of the architect. The latter narrows as the design process becomes more rigid and less iterative. An example of process standardisation enabling customisation of the product and giving voice to the users' preferences is offered by the Inhabitant-Driven Customisation Approach (IDC) (Eloy and Vermaas 2022). In an IDC test at the Siza Vieira Malagueira Complex, the building's inhabitants were able to customise their housing units through a dedicated design tool called MyChanges, developed for the purpose without jeopardising the architectural intent or overall design aesthetic. When enabling co-design, this combination of standardisation and customisation would contribute to preventing what has become known as the customisation paradox (Eloy and Vermaas 2022) while leading to a positive sociocultural impact. Standardisation and customisation of the design process reflect on the product and become apparent through a series of improvements, from architectural language to efficiency and affordability. A new architectural language might emerge because of the possibility of combination and personalisation from the client, originating bespoke combinations of standardised components and enabling variations in dimensions, materials and shapes in favour of a more flexible output. The new aesthetic would move away from the archetypical forms, typologies, materials and textures associated with standardised architecture of the 20th century.

Enhanced efficiency in the production and manufacturing processes should lead to a reduction in costs through the modularisation and optimisation of building components. This would result in a more affordable, high-quality architectural product with positive implications for the end users. Reduced costs and enhanced performance would enable low and middle-income groups to access architectural products having long-term impact on households, which would entail a reduced investment, long-term energy efficiency and ease of maintenance. Standardisation through BIM tools thus allows for further optimisation during the design process, paving the way for the creation of more efficient buildings and cutting fabrication costs through the optimisation of structural and material use, as well as operational costs through better response to environmental conditions. This can, if applied thoroughly, save energy and reduce fuel poverty where optimised design becomes accessible to all.

Impact on the Democratic Design Value

This is associated with the accessibility of the design process to different actors, including architects, IT technicians, BIM specialists and end users, as well as to interoperability. Standardisation of the process can be achieved through a BIM-shared platform and specific language. As discussed before, ISO 19650 has a rigid set of protocols that dictate how each task is performed and how each component within the digital model is designed and controlled. This application demands the involvement of highly specialised personnel who speak the ISO 19650 language and its BIM lexicon. If, on the one hand, this streamlines the process, it makes it less accessible to non-experts or non-trained specialists, hence favouring the creation of new elites. In fact, while new roles and skills are created for the purpose, existing roles whose skills are no longer relevant in the new streamlined process become obsolete. Within this perspective, accessibility and opening of the standardised process would not equate to democratisation. Furthermore, standardised BIM processes entail the creation of Industry Foundation Classes (IFC) format of files, which aims to facilitate collaboration and communication between project stakeholders. However, despite this standardisation, interoperability remains a central issue that BIM users struggle with due to discrepancies occurring while exchanging data components (Ohori et al. 2021). Executing the ISO 19650 steps remains very challenging throughout the process and requires constant care by modellers as complex elements evolve during the design process. Further to the IDC would be better to have the full name rather than IDC tools, several online design platforms started to operate in the early 2010s, such as Kujiale in China, CY Vision in the United States, Perspective in France and Varjo Technologies in Finland. These are all cloud-enabled interior design platforms that mainly focus on the home environment, offering users the possibility to scan, generate and display design solutions using Augmented Reality or Virtual Reality Are we being consistent with the use of capitals? Based on the 3D models created or adopted in the software floorplan database, users can apply various materials, furnishings or types of interior design styles. AR and VR technologies provide reality experiences for users to validate their design preferences. To use platforms like Kujiale, users do not necessarily need to have specialisation or expertise to generate design solutions, as the information and knowledge gap between design professionals and users is narrowed.

Product standardisation impacts the democratic value of design through sociocultural and economic implications. Accessibility of the product to a larger group of users, in fact, is only possible when the market price becomes affordable. Affordability, in turn, results from the contained production costs enabled by process and product standardisation combined. This is possible within the BIM workflow and in BIM 7D, which also enables building managers to have access to the BIM model during operation and facilitates maintenance procedures, allowing for better-informed retrofitting, adaptations or changes in the short and long term. The trade-off is that such efficiency would entail the involvement of specialised staff through operation and maintenance stages to access, read, translate and update the BIM model during the building's lifetime (Pinti et al. 2022).

Impact on the Environmental Design Value

The BIM landscape transforms all building components into data elements that are measurable in terms of carbon footprint. Within this context, BIM 6D incorporates environmental and life-cycle targets within the design process, optimising material selection, construction procedures and time management. For example, integration of BIM with Life Cycle Assessment (LCA) has enhanced the efficiency and accessibility of environmental impact analysis of the built asset for construction practitioners (Zheng et al. 2023). The role of BIM can also be explored in relation to Construction Waste Minimisation (CWM) with the development of dedicated plug-ins featuring layers for waste management and waste processing cycles, relying on big data technologies (Bilal et al. 2015). Furthermore, BIM allows for the detection of clashes between objects, such as Heating, Ventilation and Air Conditioning (HVAC) ducts and structural beams, or outstanding issues in the execution workflow, which requires some activities to be executed before others. Automating these coordination activities creates positive environmental impact by reducing carbon emissions, benefiting design and construction teams as well as users. However, digital technology requires the use of powerful computers and large data centres to store increasingly complex data within the BIM model. Albeit invisible, since most BIM model storage is cloud-based, these servers have an environmental impact requiring increased energy requirements, in addition to security and privacy concerns.

Impact on the Economic Design Value

The standardisation of the design process and product has the potential to substantially lower costs for both the industry and end users. Considering the economic impact created, accurate cost estimation enabled by BIM 5D results in the optimisation of costs, process time and materials. An estimated cost reduction of 15–20%, enabled by BIM, is extremely appealing to governments aiming at reducing the costs of public sector construction, as shown by the 2011 UK Government-Industry BIM programme. This promotes the adoption of BIM in public and private sector organisations involved in procurement and delivery of infrastructure (Construction Leadership Council 2013; HM Government 2012). A most recent example is offered by the 2016–2020 Government Construction Strategy, aimed at developing digital capability and embedding digital technologies to increase productivity (The Infrastructure and Projects Authority 2016). Although high-level assessments suggest a reduction in the costs of the process, it has been demonstrated that the adoption of BIM in small–medium companies can generate indirect costs for clients, which are not well received. At the corporate level, standardisation and optimisation enabled by BIM suggest an increase in profitability due to scheduling efficiency and operation cost reduction and to a better control of cash inflow and outflow patterns at all stages of the project. However, the high initial costs and low return on investment represent a barrier for the small and medium-sized enterprises, preventing widespread investments in favour of a digitalised and standardised process.

Conclusion

In 2017, the RIBA reported that two-thirds of UK practices are using BIM. The widespread deployment of BIM workflows is becoming increasingly common in design, fabrication, construction and operation (Malleson 2017). The UK Government mandated the use of BIM for all public projects starting in 2016, aimed at reducing design and production costs, achieving efficiency and being at the forefront of technological advancement within the construction industry. It was noted that:

> *Industry must therefore meet the challenge – only through the implementation of BIM will we be able to deliver more sustainable buildings, more quickly and more efficiently. BIM is also critical to the successful implementation of a wider offsite manufacturing strategy. (Construction Leadership Council 2013: 9).*

Albeit not enforced, its use and application have also been highly encouraged in EU countries (Volt and Toth 2020) and North America (U.S. General Services Administration 2022). The benefits are clear and established in each of these frameworks, increasing efficiency and establishing a common collaborative landscape for the AEC industry, which aims at saving resources, time, cost and carbon emissions. There also seems to be impetuousness towards its implementation as a positive innovation and progression in the industry. However, promoting it as a tool that would solve all design and construction problems lacks practical implementation steps and overlooks the vast investment and application procedures in making it operational and accessible to all stakeholders in the AEC industry (Jasinski 2021). Implementation remains a complex and costly task, requiring substantial changes in the industry's structure and operation. In practice, architects struggle with BIM workflows, especially in the early stages of the design process. The architectural design process is fluid, iterative and non-linear by nature (Brawne 2003), while the overall BIM language requires precision and utmost rigour given that it is based on data components rather than abstract visualisation. The inherent ambiguity of design thinking in its traditional format is almost impossible to achieve within the BIM ecosystem:

> *Designing is not a linear experience, in which you have an idea, put it down on paper, then carry it out and that's that. Rather it is a circular process: your idea is drawn up, tried out, reconsidered, and reworked, coming back again and again to the same point. As a method it seems very empirical, but if you look around, you realise that it's typical of many other disciplines: music, physics, astrophysics too (Renzo Piano in Brawne, 2003, p. 78).*

Beyond the impact on the design process, the adoption of BIM demands a shift in the process of value formulation that offers benefits for the end-user. In fact, literature shows that the benefits of using BIM praised by professionals and specialists are largely unacknowledged by clients, who refuse to bear the related costs.

> *You don't buy an iPhone because Apple has a great process for manufacturing, but for the benefits you get when you use the end product. Similarly, we have to constantly ask what the benefits of BIM are to the end customer and how these can be quantified (British Standards Institution 2021: 9).*

This highlights the need to redistribute BIM developmental and operational costs among the interested parties but also shows the lack of a perceived value from the client's perspective, which needs to be addressed by reformulating new value propositions and reviewing the companies' value chains. From this standpoint, and as BIM is slowly becoming more of a requirement from governments, developers and owners, this opens the debate around the viability of BIM workflows and ISO 19650 and the contested role of architects and other designers within the design process. As previously argued, ISO 19650 introduces a new language that supersedes the existing drawing tools, namely orthographic projection (plans, sections, and elevations), which are becoming obsolete. This will eventually change the way people practice architecture, engineering and construction, entailing a natural transformation in educational systems. Additionally, while the discussion around standardisation and customisation within the context of BIM continues, design workflows are witnessing another technological innovation, known as artificial intelligence. The latter is disrupting industries at a rapid pace and is underlining the question about the future of AEC professions. Automation in visualisation, simulation and analysis of design solutions is multiplying the possibilities in the construction industry and reinforcing its digital and information language. However, as AI tools are increasingly accessible and ubiquitous, these tools could disrupt standardised BIM workflows before they take full shape. Implementing AI within the design workflows requires a shift in the whole design thinking to adapt to this new paradigm. Carpo (2023) argues that:

> *There is therefore a certain irony in that the latest avatar of electronic technologies may now foster a revival of art theory tropes that twentieth-century modernism tried to eliminate from the visual culture of the machine-made environment. Imitation and style, which modernists had rejected from art theory, are coming back to art practice through the window of technology.*

It has been argued that the world seems to necessitate more customisation than standardisation, both in the process and product. Individualisation and differentiation are key values in today's society, its products and buildings. Simultaneously, the making process is not exclusive to experts formally educated in universities; rather, it is disrupted by self-taught entrepreneurs who can integrate new tools using artificial intelligence.

References

Allen, S. (2005). The digital complex. *The Log*, 5(Spring/Summer), 93–99.

Bilal, M., Oyedele, L.O., Qadir, J., Munir, K., Akinade, O.O., Ajayi, S.O., Alaka, H.A. and Owolabi, H.A. (2015). Analysis of critical features and evaluation of BIM software: towards a plug-in for construction waste minimisation using big data. *International Journal of Sustainable Building Technology and Urban Development*, 6(4), 211–228.

Brawne, M. (2003). *Architectural Thought: The Design Process and the Expectant Eye* (First Edition). Oxford: Architectural Press.

Brensing, C. and Schittich, C. (2013). *Building Design at Arup*. München: DETAIL. https://doi.org/10.11129/9783955530105.

British Standards Institution. (2021). BIM perspectives: unlocking value. https://www.bsigroup.com/globalassets/localfiles/en-gb/built-environment/bsi-be-bim-perspectives-case-studies.pdf (accessed 5 December 2023).

Carpo, M. (2016). Parametric notations: the birth of the non-standard. *Architectural Design,* 86(2), 24–29.

Carpo, M. (2023). Imitation games. Mario Carpo on the new humanism. *Art Forum*, 61(10), 184–188. https://www.artforum.com/features/mario-carpo-on-the-new-humanism-252735/ (accessed 28 November 2023).

Construction Leadership Council (UK Government). (2013). *Industrial Strategy: Government and Industry in Partnership: Construction 2025*. London: Crown (The National Archives).

Deutsch, R. (2015). *Data-Driven Design and Construction*. Hoboken, NJ: John Wiley and Sons.

Drew, C., Hay, B., Dudas, T. and Dewfield, S. (2022). *The Design Value Framework*. London: The Design Council.

Eid Mohamed, B. and Carbone, C. (2022). Mass customization of housing: a framework for harmonizing individual needs with factory produced housing. *Buildings,* 2022(12), 955. https://doi.org/10.3390/buildings12070955.

Eloy, S. and Vermaas, P.E. (2022). Architectural approaches to housing customization: introducing the inhabitant-driven customization approach and the MyChanges tool. *Archnet-IJAR: International Journal of Architectural Research,* 16(2), 396–412.

Hillier, B. (2007). *Space Is the Machine: A Configurational Theory of Architecture*. London: Space Syntax.

Holzer, D. (2016). *The BIM Manager's Handbook: Guidance for Professions in Architecture, Engineering and Construction*. Chichester: John Wiley and Sons.

HM Government. (2012). *Building Information Modelling*. London: Crown (UK Government).

International Standards Organisation – ISO. (2023). ISO. www.iso.org (accessed 28 November 2023).

Jasinski, A. (2021). Impact of BIM implementation on architectural practice. *Architectural Engineering and Design Management,* 17(5–6), 447–457.

Kuma, K. (2009). *Studies in Organic*. Tokyo: Toto Publishing.

Malleson, A. (2017). Two-thirds of practices are now using BIM. *RIBA Journal*, 5 July.

Ohori, K.A., Krijnen, T. and Stoter, J. (2021). An inspection of IFC models from practice. *Applied Sciences,* 11(5), 2232.

Pinti, L., Codinhoto, R. and Bonelli, S. (2022). A review of building information modelling (BIM) for facility management (FM): implementation in public organisations. *Applied Sciences,* 12(3), 1540.

Piroozfar, P.A.E. and Piller, F.T. (2013). Mass customisation in environment, architecture and construction: an introduction. In P.A.E. Piroozfar and F.T. Piller (eds.) *Mass Customisation and Personalisation in Architecture and Construction*. London: Routledge. 3–14.

Schumacher, P. (2008). Parametricism as style – parametricist manifesto. https://www.patrikschumacher.com/Texts/Parametricism%20as%20Style.htm (accessed 28 November 2023).

The Infrastructure and Projects Authority. (2016). *Government Construction Strategy 2016–2020*. London: Crown (UK Government).

U.S. General Services Administration. (2022). 3D-4D building information modeling. https://www.gsa.gov/real-estate/design-and-construction/3d4d-building-information-modeling (accessed 1 December 2023).

Volt, J. and Toth, Z. (2020). *EU Framework for Buildings' Digital Logbook*. Brussels: European Commission.

Zheng, B., Hussain, M., Yang, Y., Chan, A.P. and Chi, H.L. (2023). Trade-offs between accuracy and efficiency in BIM-LCA integration. *Engineering Construction and Architectural Management*. https://dx.doi.org/10.1108/ECAM-03-2023-0270.

9

Data, Information or Design: New Ways of Conceiving the Value of Architecture in the Global, Digital Age

Urša Komac[1], William McKee[2], and Robert Osei-Kyei[1]

[1] *University of Western Sydney, Australia*
[2] *University of Sydney, Australia*

Introduction

Architecture represents the aspirations of people living in a particular era (van der Rohe 1924). In 1485, Alberti's book De re ædificatoria, which translates as On the Art of Building, clearly delineated the territory of the architect and the builder. The territory of the architect is to generate a special building (Alberti 1955). This edifice is to righteously judge what fits and is right. Alberti (1485) describes the business of the architect as only of extensive genius and establishes the architect's role as being that of representation of buildings, not their execution. This paradigm still underwrites much contemporary understanding of the profession. The territory of the builder is that of execution, delivery, matter and affordability *to run up anything that is immediately necessary for any particular purpose and about which there is no doubt of what sort it should be or of the ability of the owner to afford it. . .* (Alberti 1955: 205). This is the practical delivery and technical assembly of materials to meet the real-world parameters that enable buildings to come into being, i.e. money, time and effectiveness. As Alberti distinguishes between the two principal actors; architect and builder, it might be helpful to seek to distinguish between the essence of architecture and building. As both are nouns and vocations, they may be regarded as distinct. Sometimes, but not always, one entity can embody both, and a building can be viewed as architecture and architecture as building. For the purposes of this investigation into the impact of the digital, it is more useful to consider them separately. The issues will be examined to affirm that there is indeed well-founded belief that a difference exists. To begin, it needs to be asserted as to what defines architecture. Moreover, how architecture is relative to or even contrary to building is significant. People are surrounded by built

Architectural Design and Management in the Digital Age: International Perspectives, First Edition.
Edited by Bob Giddings, Silvio Melhado, and Jenni Barrett.

forms – cities, towns, villages, bridges, roads, temples, houses, etc. An estimated 5% of single residential dwellings in Australia are designed by an architect (AACA 2023). If the involvement of an architect results in architecture, it might be that only 5% of single dwellings is considered to be architecture. Therefore, the remaining 95% are mere buildings. The deficiency of these buildings should be identified that determines why they are not described as architecture, as well as the demonstrable virtue of the 5% that enables them to be designated in that way.

Building has undeniably been improved because of digital technologies. Building Information Modelling, as an example of a digital technology used in Australia and New Zealand is cited particularly for reducing errors and omissions by 45% as a result of integrated and collaborative 3D modelling capacity (McGraw-Hill Construction 2014). In 2011, 1 Bligh Street in Sydney achieved Australia's highest green building rating (GBCA 2023). This sustainability credential was attained through the essential performance testing and simulation of the building by BIM software (Financial Review 2011). BIM was also found in a 2012 study to account for a 37% shortening of a building process compared to a traditional delivery (McGraw-Hill Construction 2012). Conceivably, this equates to a greater capacity to deliver buildings at greater speed, which has direct implications regarding reduced cost. The data supports a case for higher quality against a metric of cost, performance and speed for a digital tool's transformative capacity on building as process and product. Yet, it needs to be considered as to whether BIM has added measurable value to architecture. The Villa Savoye in Poissy, near Paris, is one of the most iconic and lauded examples of architecture from the Modernist Movement, but it leaked. In June 1930, Mme Savoye, patron of the modernist masterpiece, wrote to Le Corbusier that water is still leaking into the garage (Sbriglio 2008). It therefore needs to reflect on how and why a building with such fundamental failings as a roof that cannot keep the rain out should be considered a success. The reason is because the Villa Savoye is not measured against the metrics of performance, but rather it is measured against the criteria of architecture as defined by Le Corbusier himself, in that *Architecture is the learned game, correct and magnificent, of forms assembled in light* (Le Corbusier 1927: 37). Mies van der Rohe had his own struggles with an unhappy client, complaining of cold and comparing a stay in her commissioned architectural marvel to camping. Were the Farnsworth House to be measured and evaluated as a building, then likely it would be found wanting. The heroes of modernism often define architecture as something that grapples with larger powers. Statements by Le Corbusier and Mies van der Rohe are typical in their obtusity, loftiness and vagueness. It is perhaps spiritual, reaching for something beyond utility, stretching towards meaning. As such, while it might be quick and easy to affirm the impact of the digital upon building in the positive sense, the task of considering the impact of the digital upon architecture is altogether more complex. What made architecture difficult in the 15th to the 20th centuries remains the same in the 21st century. It is how to imbue spaces with meaning that transcends the utilitarian provision. The focus of this text, therefore, is not to evaluate whether the digital toolkit has assisted in making higher profits and more functional buildings, as they have been achieved. Rather, is whether faster, cheaper and better-performing buildings and building processes equate to better architecture.

Computer-aided Design – Harbinger of the Digital–Human Workforce

In October 1960, Ross (1960) used the acronym CAD to describe computer-aided design for the first time in a technical memorandum to describe a project undertaken by the Computer Applications Group for which he was the head project engineer. *The objective of the project was to evolve a man-machine system which will permit the human designer and the computer to work together on creative design problems* (Ross 1960: Abstract iii). This twenty-two-word quotation introduces CAD as the vanguard of digital technologies in the architectural process, heralding what would become a hybrid workforce and explicitly describing the man–machine relationship. Furthermore, we can infer a required amount of communication between man and machine within this relationship. Communication requires language, and as each digital machine and tool develops its own language, the notion of translation needs to be addressed. Digital tools should be explored in terms of language. Each tool has within it a language to interface with a tool, and an operator must learn to speak that language. It therefore follows that, as each digital tool has its own language, it is necessary for an operator to speak multiple languages and to be a digital multi-linguist capable of not only understanding the source information but also translating that information to another language so that other digital tools can process the data. Translation from digital tool to digital tool is reliant on a preceding exchange between people and machines, whereby operators can communicate the initial human information to the machine in a way that the machine is capable of understanding. So the principles are that data populates the digital architectural environment, which is represented as tools, and architects use these tools in the way that perhaps a craftsperson would have done (Sennett 2009). The scenario is therefore quite a complicated situation, especially in terms of roles and communication. Architects are tasked with working their way through the design process but may not be familiar with the required languages of the different tools. Therefore, operators are required to intervene to achieve sufficient quality of communication between the tools, as software interoperability is far away from ensuring it. There are clear risks in fragmenting the process in this way, as architects and operators may not have a common empathy.

Instructive or Questioning Communication

An important actor within the field of architecture and digital development is the architect Frank Gehry, an early adopter of the digital modelling software CATIA (Lindsey 2001). Gehry Technology developed and patented a particular architectural process focused on a simplified version of the software, Digital Project, to produce their information for building designs (Digital Project 2023). It can be deduced from the velvet and wax models for the Lewis Residence (Kipnis et al. 2003), and in fact, from the company's ongoing insistence on paper and card models, that for Gehry, the computer is a necessary tool for delivering forms already designed. Gehry's sculptural and exuberant forms simply demand more advanced digital tools with which the designs can be realised. This kind of communication can be characterised as instructive, which is where an intention exists in the human mind and the machine is fed instructions to execute this intention. It is not important for the human to understand the language of the machine. It is only important for the machine to understand the human language, a unilateral, not reciprocal, hierarchal form of communication. An alternative communication between man and machine may be referred to as questioning. Rather than instruct, the human actor poses a question. An example is Greg Lynn's Embryological House (Lynn 2013), whereby a set of parameters is defined and the question is posed to the machine for it to answer. This is a form of iteration or generative design whereby the question defines a scope and the machine provides suggested answers, which fall within that scope. As a communication, it has a certain degree of reciprocity, and the output may not be known or predetermined at the beginning of the exchange. If this is mapped onto the initial analogy, then this would play out as a craftsperson exerting an intention upon the tool, and when the tool encounters a material defect, the craftsperson allows the original intention to be reviewed, enabling the tool to suggest alternative outputs that may differ from the original intention.

A further development in communication could be that of an intimate and constant exchange, consisting of an ongoing process of instruction, questioning and suggestion, being offered and generated by both man and machine. Such a process would be a flat hierarchal structure that represents a truly exciting aspect of the man–machine dialogue, whereby the machine becomes creative equal and co-author of the product. It represents a final illustration in the craftsperson analogy as a fully symbiotic relationship, whereby both human and tool are in constant dialogue so that the output is a product of both tool and craftsperson. This idea is by no means new. The discussion has existed throughout the ages about materials. The idea can be viewed as the expression of recognisable and naturally occurring characteristics of a material; for instance with timber, there will be grain, knots and other imperfections. Even more than its tactile or visual characteristics, timber as a material has an appropriate set of tools dedicated to it, including chisels, hammers, nails and sandpaper. Frank Lloyd Wright sought to explore and express the innate characteristics of concrete—a pliable material that does not necessarily have an innate characteristic form. He sought then to impose or extract its quality to form a physical image (Wright 1928). This illustration could be extended to Tadao Ando and the presence of leaf imprints in poured concrete for his 1993 Conference Pavilion at Vitra. Two leaf imprints sit next to a hole remaining from the tools of concrete construction, generated by the formwork. The presence of this leaf was not predetermined, it was the product of a process. The hole in the formwork and the leaf are graphic indicators of the nature of casting. The craftsmanship is not the premeditated inclusion of the leaf or the hole but rather the decision to not conceal them in the final appearance. This represents a deep engagement with the tools of production and a dialogue that permits and embraces suggestions, errors or process indicators. In the case of Gehry and the Lewis Residence, Lynn's Embryological Houses and Ando's Vitra Pavilion, there is a deep recognition and critical engagement with the production tools. It raises the question of how these ideas can be provocative within digital tools and architectural production. Digital tools permit Gehry to deliver buildings that otherwise would be unachievable within the parameters required for feasibility. For him, the digital is the means and not the end. However, the digital is not the reason that Gehry is an internationally recognised and undeniably important architect. The reason is that he has willed the digital toolkit to achieve an architectural aspiration. Lynn's built work by comparison is more modest in scale and quantity but uses digital tools to pursue an architectural principle. The digital is the material with which Lynn is working, rather than clay or plastic, and this is pushing the Overton window, i.e. the political possibility defined as the range of ideas that the public is willing to consider and accept (Conceptually 2023) for architectural form and architectural theory.

Today, CAD as a term has fully pervaded the public imagination and exists within the public vocabulary. New students of architecture invariably come to their education with some pre-conceived notion of CAD, endowing their learning of it with a special importance, a significant milestone in their budding education. Likewise, architectural clients often have an inkling of CAD and its role within the production of the information they are presented with. Students and clients, as well as the public, feel that architects sit at computers, and with a few clicks, architecture comes out. There is a further idea, which is that this powerful tool has replaced much of the work of the architect and is capable, at infinite speed and minimal human effort, of almost instantly producing drawings, models, design options, etc. Of course, this is not true. This myth of the digital extends beyond CAD and its descendant technologies, such as BIM and 3D modelling, to the entire digital toolkit. It centres around a disproportionate sense of speed and a fabled-go button. As with most myths, it is not born out of nothing; there is of course a grain of truth, but the extent to which the duties of the human have been alleviated is exaggerated. Time remains a non-negotiable input requirement, regardless of the digital tools used in their production. To refer to the objective set out by Ross (1960), the digital production still requires man and machine to work together.

Daniel Cardoso Llach states that *technologists attempt to codify the manual craft of operating a milling machine [. . .] instead of reducing work, transformed it [work] and relocated it from the shop to the programmer's desk* (Llach 2015: 50). This quotation illustrates a very early example of an automating machine that, at scale, reduces time but in fact relocates that time to a different stage of the process. The go button of the digital myth may seem to exist but only by virtue of a great deal of time that has been invested elsewhere in the process. As with all things, the greater or lesser input of human time invested has a direct impact upon the quality of the output, whether this be drawing, modelling, 3D simulation or building. According to Sennett (2009) it is the craftsman who engages with the process. The tools themselves are not magical, and a good output is reliant upon a human capacity to integrate, invest time and most importantly, direct the activity. The tools used will exert influence on the user and the work. The proposition is that, as with the approach of the craftsperson, a human may be conscious of this influence and critically evaluate it. These tools require an understanding of how they work and whether their mark can be read in the drawings, buildings and architecture they produce. The question thus arises as to whether digital tools such as ArchiCAD (Graphisoft 2023), for example, have a distinctive visual maker's mark embedded in its products (Lawson 2002). This mark could be present within either the documentation or even in the appearance of the buildings it produces. It might be that a building design produced on Rhinoceros (McNeel 2023) looks different to a building design developed on ArchiCAD , but nevertheless all Rhinoceros buildings can appear similar. It might be perceived that recognising the package on which a building was produced is an indicator of lack of skill when a tool of fabrication can be detected. The motive behind obscuring or revealing the tools of production lies in how original or bespoke each design should be. The question of the maker's mark and decisions of how much influence the machine is allowed and how much of the innate characteristics of tool or material are legible is really a question of creativity. Neither revealing nor concealing is possible in isolation, but craftsmanship is perhaps the approach that guarantees whether to conceal or reveal has been chosen. The revelation or concealment is purposeful. It might be that a beautiful image could be created in Microsoft Paint or, alternatively, in a clear and informative drawing set offered by Revit (Autodesk 2023). Each tool could have a specific purpose. Thus, it may be a complete mismatch if it is attempted to use a digital tool made for one purpose for another purpose. A common British axiom is that a bad workman always blames his tools. The inference being that the workman is bad and that the tools are inanimate, the work would have been bad regardless of the tool. However, this distinction blurs where digital tools are concerned. Architects and operators assume that their mastery of increasingly intelligent and complex digital tools means that they have total control. Yet, especially with the growing influence of artificial intelligence, it might be that the tools are guiding the designers. A more detailed comparison of the two software programmes already mentioned is relevant here. As a digital tool, ArchiCAD is quite prescriptive, precise and immediately conjures reality. For instance, it has windows made of glass and frames that fit into a process of assembly. If the architect or operator does not possess sufficient knowledge and experience, they are to some extent led by the tool. In contrast, Rhinoceros is a modelling tool that deals in pure geometry – points, curves and XYZ coordinates. There is no such component as a window in the language of Rhinoceros, only a surface and a hole in that surface. Rhinoceros as a digital tool does not differentiate between a building, a cup or a character in a game. It is a series of cartesian relationships. As with a hammer or pencil in the shed of the craftsperson, the digital craftsperson must learn the general appropriateness of their tools and then develop a particular repertoire suitable to their own creative endeavours. Kahn (1982) discussed with his bricks, as to what they wanted to be, and a similar situation may arise with computational and digital tools.

To consider digital tools in relation to the craftsperson raises the question of the digital environment where these tools are put into operation. A digital environment can be understood as a digital space or interface with a particular tool. This definition can be expanded to include the computational hardware and the wider digital infrastructure requirements for a tool to be operated effectively. In the case of rendering software to produce high-resolution imagery or simulation software that can model environmental loads and conditions, these digital tools also come with hardware requirements – memory, graphics performance, stable internet connection, etc. From the digital environment, this can infer economic contexts, geographic contexts and infrastructural contexts. If, as has been previously suggested, digital tools leave marks in the outputs they produce, then a similar impact must be attributed to the environment in which these tools are deployed. The extent to which the digital environment or context impacts the output and subsequent building is a question that has to date received little consideration. Such a digital environment could literally be a geographic location, which is closely aligned to how vernacular interpretations are categorised. Aligning, or seeing similarities according to groupings related to geography, could similarly be categorised based on computational metrics, a kind of techno-regionalism or techno-vernacular. As opposed to an architecture being born out of a geographic, temporal or geological context, it is born out of a technological one. Such categorisation invariably leads to seeking identification and extracting characteristics from various digital contexts. Llach (2015) discusses the weight of a model in terms of its digital size. He considers how big the file is, what that indicates and what the impact might be. Digital space is imagined as being free from physical restraints. For example, a digitally extruded beam can theoretically span an infinite space, but perhaps a model is still restricted by alternative parameters such as computational power. Insufficient computational power will limit the software's ability to run effectively. In turn, this impacts the type and style of image that can be produced and, ultimately, the building outcome. It is not unreasonable to consider that a building's image or complexity could be determined by the capacity of the machine on which it is produced. Thus, the digital craftsperson is in constant dialogue with not only the tools but also materials and the environment. This is analogous with the software becoming the tool, the digital model becoming the material and the computing hardware and digital infrastructure being the environment. In a damp climate, timber will swell. With limited memory, a computer programme will lag. This environmental impact, whether real or digital, is accommodated by the adept craftsperson. In a digital environment, operations are undertaken sequentially to minimise computational load, and a wire-frame view could therefore be selected. This envisages an architecture that speaks of not only the digital tools used for its development or documentation but also the digital environment in which it was made. As Carpo (2011: 88) notes that *buildings remain to this day a hybrid of hand-made and machine-made parts and operations; partly ready-made and mass-produced, partly custom-made and made on demand, or mechanical prefabricated but manually assembled on site.*

Hybridity

Carpo (2013: 8) points out that *a meaningful building of the digital age is not just any building that was designed and built using digital tools: it is one that could not have been designed or built without them.* The idea of hybridity is central to the preceding discussion of language and the nature of the relationship between people and machines. Language is the reason for communication between humans and machines. However, something is always lost in translation. When documentation for a building is undertaken by hand, the level of precision is limited by the width of the lead of a pencil. Scale is deployed to allow the human to show greater detail, but the actual line always has a real-world thickness. As a result, buildings of the predigital era contained an inherent level of imprecision. Yet, numbers of materials and their construction is not very precise either. From inception to completion, the information passed through human hands, and details were mis-communicated often due to their two-dimensional representation. Ideas of exactitude varied, but the basic instrument of human evaluation of information was assured. With the advent of digital methods, human-to-human passage is much reduced. More accurately, there are a series of exchanges from human-to-machine-to-human-to-human. This sequential translation reveals certain differences in the fundamental capacities and priorities of engagement and instruments of evaluation between humans and machines. A digital tool for modelling can work to decimals of a millimetre. Though the human eye could not perceive a fraction of a millimetre across a span of metres in the real world, in the virtual world, this fractional discrepancy will reveal itself every time a measure is typed for a component. A single discrepancy can compound across a model or drawing sets. It can become a time-sucking and mind-numbing process of finding the original error and amending it, all of which is immediately moot when translated to a building site and human contractor. This introduces the notion of abstraction of digital tools from reality and the importance of developing specific tolerances for particular materials and construction. It is important to set the tolerances at the beginning of the process so that they can be checked at all stages, and time is not spent adjusting discrepancies that are well within the tolerances, as there is no value in such an extremely accurate point cloud on a survey. Perhaps it can be understood as some kind of simulacrum. The capacity of the computational world has become more powerful than the reality perceived by the human instrument. Reality is being asked to contort itself to reach a standard that the computer is forcing us to believe is normal. In terms of pressure, this is substantial. The capacity of machines to measure is forcing an increased level of precision on people. For users, the capacity of the computer to operate at 0.001 of a millimetre is of little value, as the humans cannot perceive or evaluate this. If the wider digital environment is considered, those concerned with energy and limiting environmental impact might welcome a greater level of accuracy. Yet, even in these forms of engineering, performance levels do not match the precision of digital modelling.

So far, the craftsperson has been lauded as having a virtuous or positive role. However, as the embodiment of human values, the craftsperson could detrimentally restrict the full capacity of the machine and the digital toolkit to improve the built environment. The craftsman has an intensely human persona. To deploy this term within a digital discussion is to encourage a deep and critical engagement with digital technologies to achieve human ends. This comes back to the question of definition, i.e. architecture versus building. If it is believed that architecture can be reduced to a discrete set of parameters, then possibly a point might be reached where computers can produce architecture at a faster speed, greater efficiency and more innovative level than humans. Nevertheless, if it is posited that architecture is too complex to be discreetly defined, or that architecture contains an aspect of spiritual aspiration that can only be understood by the designers, then digital tools will always require human input, and the process will remain hybrid. What an architect must do, or indeed any designer must do, is to consider what they wish their role to be within the unfolding hybrid workplace of people and machines. The virtue of a craftsperson approach has been discussed, and the ephemeral contribution that a craftsperson brings is understood. In terms of m^2, weight, cost and quantity, the machine can and will usurp human capacity, but the amenity attributes are much more difficult to measure. Inevitably, the idea of evaluating architecture within a digital context requires a more clear definition of what the objective of architectural endeavour might be. The power of Alpha Zero (Somers 2018) was unlocked because the objective of chess is quite simple. To win, you must establish checkmate. When measured against a metric such as a win rate, AlphaZero could be deemed to be extremely successful. Yet, more subjective evaluation criteria would be necessary if other attributes, such as beauty, were involved. A problem with this issue is that architecture does not have a consensus of agreement about what beauty is. Within the context of a climate emergency, energy prices or affordability, the digital toolkit can be of enormous value. It can generate data, process iterations, and look at optimisation models as a means of successfully meeting one or more parameters. It is also recognised that there is enormous potential within machine-to-machine production, whereby a design machine develops and passes design instructions to a construction or fabrication machine. This mode of production would no doubt result in levels of accuracy, efficiency and precision previously unachieved by human action. Nonetheless, this machine-to-machine chain of interactions eventually encounters human beings, as the overarching idea of architecture is that of human occupation. As such, the digital environment or toolkit will eventually be evaluated by humans using their senses as the instruments of the evaluation.

If beauty is used as a surrogate for architecture, there should be a discussion as to whether it could be applied to a building, which is the product of a pure computation situation with its logic-solving capabilities. It cannot be ruled out but seems unlikely. The creative process is difficult because it is full of contradictions, backward steps, happy mistakes, compromise, restriction, innovation, limitation and so on. Beauty may be understood as *the harmony and concord of all the parts achieved in such a manner that nothing could be added or taken away or altered except for the worse* (Alberti 1955: 115). While it is important to seek beauty in the ability of humans to deploy the new digital toolkit, the difficulties remain the same. The objectives of creating something beautiful remain just as vague, ambiguous and evasive. There are larger toolkits with improved capacity to extract data, simulate performance and integrate information. All seem to be aimed at generating buildings as opposed to architecture, and the latter remains elusive. Anyone who is an educator will at some stage doubtless have been asked a question of penetrating simplicity to distinguish between architecture and building. The distinction is a degree of criticality, offering two available routes. The first is based on Le Corbusier's definition of the house as a machine for living in. It could be argued that if it is a machine, maybe it should be the product of a machine. Digital advancements in material technologies, data extraction, dissemination of information, advancements in global communication networks and peer-to-peer systems could be harvested as a resource with which to iterate and refine the house as a high-performance instrument. The second is the house as refuge, connected to nature, constructed of biophilic materials, responding to the seasons and protecting from an increasingly threatening climate. This requires a level of sensitivity and understanding that is difficult to envisage ever being produced by a machine. Beauty, cultural identity, functionality, efficiency and usability are uniquely human traits. The human-to-human quality of architecture, regardless of the means of production, is perhaps unique, too individual and too elusive to be computationally created.

Conclusion

It is still a gestation period where digital tools are largely being asked or forced to undertake activities currently being done by people. There is still a hangover from a global pandemic, there is war in Europe, inequality is rife and there is a housing crisis that calls for increased volume and affordability. Overlaid upon these issues is the threat of climate change and the more global efforts to shift to greener production of energy. The pressure to increasingly offload more responsibility for the procurement and production of architecture to the machine is both enticing and concerning. Thinking back to 1951, it was the year in which the Farnsworth House was completed. It was a time of significant technological transformation.

It might be felt that Mies was pre-empting a mechanisation of the process of architecture. It could be debated if Mies was constructing a purposefully obtuse job description to indemnify him and his profession against caving into tasks that could be sequentially completed on the architectural factory line. In the 21st century, there is another industrial revolution, one of a rapidly advancing technological capacity and fear that a machine might be better equipped to do the work than people are. Digital tools have increased the speed and level of detail with which a building can be documented and permit a level of simulation and analysis previously unprecedented, to permit an incredible level of pre-construction modelling. The essence is whether it translates into increased value in architecture. Globally, there is huge awareness that the cost of the previous industrial revolution and the extraction and consumption of fossil fuels have had a negative impact on the climate and that the construction industry has contributed significantly to it. Efforts to reduce carbon emissions in construction have had a beneficial impact on the environment while simultaneously improving the amenity of the occupants. Architects should be able to assist in producing models of more affordable housing and the development of high-performance new and existing housing. Integrated into the architectural toolkit are technologies that increase speed and accuracy and allow for multiple modelling and simulating of economic, energy and efficiency data. This is clearly beneficial but does not address the fundamentals of home.

Architecture is essentially about humans rather than buildings. The latter is the mechanism by which hypotheses about the human condition are constructed, projected and tested. Buildings could be perceived as the instruments but not as music. They need to be manipulated to achieve a higher ideal. Architecture is at the top of the hierarchy. If it is asked how and why digital technologies have added value to buildings, then it is a relatively straightforward response. The more difficult question is to consider whether the digital toolkit can be deployed to impact the top of the hierarchy, to pursue human ideals and to alleviate or enhance the human condition. An alternative proposition is that digital technologies are detrimental to architects' pursuit of those goals. It may be that the digital toolkit is devaluing the core idea that sits at the genesis of architectural representation. Digital technology may produce solutions too quickly and modify them too easily. It could be that the knowledge of solutions being readily changed may impact the level of thought put into their generation in the first instance. If these tools are really saving time, it could be investigated to see where the time saved is being otherwise allocated. Alberti (1955) discusses the value of time for review. It is indeed questionable whether the time saved is being used for the much-needed task of reviewing. If this were the case, it would be possible to hypothesise that the digital toolkit could in fact assist in creating better architecture by permitting time to undertake that necessary activity. It is more attainable beauty that must be demonstrated for a building to be considered architecture. Architecture is art at its core. As such, architecture's value remains in its capacity to communicate meaning that cannot be measured by an objective instrument but is the province of the human body, mind and spirit. As such, the principal discussion around digital technology should be about its capacity to improve architects' ability to create meaning. The transition from analogue to digital, as the term transition suggests, is full of demands for different kinds of data as a means of improving the information available to designers and managers. Yet, there has been little assessment of whether all the inputs to the design process can be digitised in this way, and whether valuable nuances might be excluded. The challenge is how to engender and balance – beauty, cultural identity, functionality, efficiency and usability within the digital environment.

References

Alberti, L. B. (1955). De re ædificatoria. In L.B. Alberti (ed.) *Ten Books on Architecture*. London: A. Tiranti.

Somers, J. (2018). How the artificial-intelligence program AlphaZero mastered its games. *The New Yorker*. https://www.newyorker.com/science/elements/how-the-artificial-intelligence-program-alphazero-mastered-its-games (accessed 30 December 2023).

Architects Accreditation Council of Australia – AACA. (2023). Industry profile. https://www.aaca.org.au/wp-content/uploads/Industry-Profile.pdf (accessed 30 December 2023).

Autodesk. (2023). Autodesk revit: BIM software to design and make anything. https://www.autodesk.com.au/products/revit/architecture (accessed 30 December 2023).

Carpo, M. (2011). The craftsman and the curator. *Perspecta*, 44, 86–200.

Carpo, M. (2013). Introduction-twenty years of digital design. In M. Carpo (ed.) *The Digital Turn in Architecture 1992–2012* Vol. 8. New Jersey: John Wiley & Sons.

Conceptually. (2023). The Overton window of political possibility is defined as the range of ideas the public is willing to consider and accept. https://conceptually.org/concepts/overton-window (accessed 30 December 2023).

Digital Project. (2023). www.digitalproject3d.com (accessed 30 December 2023).

Financial Review. (2011). Raving: Bligh's crew revels in the BIM revolution. 28 April 2011. https://www.afr.com/companies/infrastructure/raving-bligh-s-crew-revels-in-the-bim-revolution-20110428-ihrtv (accessed 30 December 2023).

GBCA. (2023). The rating achieved was 6 Star Green Star. https://www.gbca.org.au/docs/walks/1%20Bligh%20Street.pdf (accessed 30 December 2023).

Graphisoft. (2023). Design and deliver great buildings. https://graphisoft.com/uk/solutions/archicad (accessed 30 December 2023).

Kahn, L. (1982). 1973: Brooklyn, New York. *Perspecta*, 19: 88–100.

Kipnis, J., Ball, T. and Neff, B. (2003). A constructive madness Frank Gehry's house for Peter Lewis. Documentary. https://www.youtube.com/watch?v=OMrptF9HoAk (accessed 30 December 2023).

Lawson, B. (2002). CAD and creativity: does the computer really help? *Leonardo*, 35(3), 327–331. https://doi.org/10.1162/002409402760105361.

Le Corbusier. (1927). *Towards a New Architecture* (Vers Une Architecture) J. Rodker (trans.). Dover Publications.

Lindsey, B. (2001). *Digital Gehry: Material Resistance, Digital Construction*. Basel: Birkhäuser.

Llach, D.C. (2015). Software comes to matter: toward a material history of computational design. *Design Issues*, 31(3), 41–54. https://doi.org/10.1162/DESI_a_00337.

Lynn, G. (2013). Embryologic houses (2000). In M. Carpo (ed.) *The Digital Turn in Architecture 1992–2012*. New Jersey: John Wiley & Sons. 124–130.

McGraw-Hill Construction. (2012). The business value of BIM in North America. https://damassets.autodesk.net/content/dam/autodesk/www/solutions/building-information-modeling/bim-value/mhc-business-value-of-bim-in-north-america.pdf (accessed 30 December 2023).

McGraw-Hill Construction. (2014). The business value of BIM in Australia and New Zealand. https://download.autodesk.com/temp/pdf/mcgraw_hill_business_value_of_bim_anz.pdf (accessed 30 December 2023).

McNeel. (2023). New in Rhino 8. https://www.rhino3d.com/en/8/new/ (accessed 30 December 2023).

Ross, D.T. (1960). *Computer-Aided Design: A Statement of Objectives*. Cambridge, MA: MIT Electronic Systems Laboratory.

Sbriglio, J. (2008). *Le Corbusier. The Villa Savoye*. Basel: Birkhäuser.

Sennett, R. (2009). *The Craftsman*. New Haven: Yale University Press.

van der Rohe, L.M. (1924). Building art and the will of the Epoch!. *Der Querschnitt*, 4, 31–32.

Wright, F.L. (1928). In the cause of architecture, VII: the meaning of materials – concrete. *Architectural Record*, 63, 98–104.

Part V

Managing Digital Projects

10

The Architect's Guide to Running a Digital Job

Marina Machado[1] *and Mary Hardie*[2]

[1] *Hilti, Den Haag, The Netherlands*
[2] *Western Sydney University, Australia*

Introduction

The field of architecture, with its intrinsic fusion of design and construction, has historically positioned architects as master builders, overseeing projects from conception to completion. The traditional master builder architect was present on site throughout the construction process. Any design or material issues that arose during construction could be managed with the full knowledge of their impact on the original design intent. However, technological advancements and shifts in project delivery methods have significantly redefined the architect's role. This transformation, explored through the lens of historical evolution and contemporary dynamics, underscores the pivotal influence of digital innovation (Burr 2011). A comprehensive discussion is necessary of how changes to traditional processes are engaging architects with a digital plan of work. This framework is designed to facilitate informed and consistent decision-making, offering architects a systematic approach to navigate the complexities of the modern architectural landscape. It is argued that digital technology makes a physical presence on site unnecessary for much of the time. Local-to-international collaboration, empowered by modern technologies, is leading architects to engage in various tasks such as modelling, detailing, consulting, managing and reviewing. Digital processes can provide transparent definitions of agreed deliverables from each contributing party at every work stage. This approach ensures that each party understands their shared obligations, fostering effective collaboration and minimising misunderstandings. It enables the digital workflows from design into prefabrication, leading to a transformation in the construction projects. The efficiencies generated by this collaboration should lead to improved project outcomes for all the stakeholders involved, whether they are internal or external to the project team. To make this goal achievable, the format of information required to deliver each design stage must undergo constant review. Digital tools offer transparent and efficient alternatives, progressively replacing traditional methods of communication,

Architectural Design and Management in the Digital Age: International Perspectives, First Edition.
Edited by Bob Giddings, Silvio Melhado, and Jenni Barrett.

especially with external stakeholders who may not have experience in visualising physical reality at the design stage.

Nevertheless, as these tools evolve, their sophisticated interactions present new challenges. Just like any computer software, the anticipation of the next version is constant, demanding repeated investment, questioning compatibility with previous versions and requiring a review of sufficient system speeds and connectivity. These challenges are particularly pronounced in developing countries where resources and access to technology may be limited. They are also challenging for smaller businesses in the developed world because of the renewed pressure to have the most up-to-date software. The potential gains must be weighed against the upfront costs of innovative technology. Equity issues generated by technological advantages are not easy to solve. The balance between achieving greater capacity to deliver sophisticated design parameters and inclusive participation of all relevant project stakeholders is a difficult one to manage. It would be unfortunate if the first world drowns out the voices of others, simply because of the cost of digital tools and the training to use them. This would be a new colonialism rather than an advancement for humankind. In summary, the evolution of architecture from master builders to information masters reflects the profound impact of accelerating digital innovation. The integration of technology, standards and collaborative environments, coupled with data-driven design, holds the promise of a more efficient and innovative architectural practice, irrespective of global location or economic status. Yet, safeguards are needed to ensure that these gains are equitably spread across all the world's nations.

Virtual Master Builders: Embracing Digital Transformation in Modern Architecture

Architecture, at its core, is a discipline that combines design and construction to create functional and aesthetically pleasing structures. Dunn (2012) states that architecture consists of design and making. However, the evolution of the architectural profession has witnessed a significant shift in the roles and responsibilities of architects, partially due to advancements in technology as well as due to changes in project delivery methods. In the pre-Renaissance era, individual architects were not only the visionary designers, but many of them also held the primary responsibility for turning their visions into reality. Such individuals can rightly be described as the master builders. Their approach was necessarily holistic, and they were intimately involved in the entire building process. A few sketches and models were often sufficient to communicate the essence of a project. Some architects acted as both the creators and builders, ensuring that the design was faithfully executed on site (Garber 2014). Over time, this dynamic has progressively changed. The architect's role shifted from the holistic design and making approach, to primarily that of only design. The separation of design and construction processes became more pronounced. Construction tasks started to be carried out by contractors, and the focus shifted from architects supervising every aspect of the build to producing detailed drawings and specifications, which served as contractual documents to specify the construction. This transformation in the architect's role gave rise to a new relationship between architects and contractors. The shift towards separation and specialisation created a more adversarial atmosphere, with each party aiming to protect their interests and mitigate their liabilities. Self-interest and self-preservation could sometimes overwhelm the interests of users. Fragmentation of the design and building processes has posed several challenges, primarily in terms of communication and coordination among various stakeholders. Different professions working in their own expertise and lacking a holistic view of the design-build process could result in inefficiencies, misunderstandings and sometimes even conflicts during the construction project delivery (Asgari and Noorzai 2023; Burry 2005; Forgues et al. 2009). Using traditional design methods, where each designer works in their own silo, makes design management a difficult task. Often, clashes between building systems are overlooked during the design phase and left to be solved on site. The architect is unlikely to be present on-site full-time, so the resolution may be made by others. Furthermore, many building elements are not even designed before construction commences. However, it should be examined a little to understand why this occurs. The authority to make decisions about the design and approve the proposals can be complicated. In these circumstances, certain parties to the process can pay scant attention to the stages of design. During the latter stages, they can suddenly start taking interest and begin to make changes. Coupled with this situation, there is often a rush to start on site. This truncates the design process, and results in ill-resolved elements being constructed. The more complex the building, the more crucial the coordination of building elements may be. Figure 10.1 shows a pharmaceutical plant. Although a very simple building envelope is required, the machinery and building installation systems are heavy and complex, requiring elaborated supporting systems that may not be considered in early design stages. This can be due to engaging consultants for partial services, rather than for the whole design period. Overlooking the overall coordination of building elements increases the risk of costly re-work and delays during construction stages (Krönert 2021).

The use of digital tools and simulations has become a cornerstone of modern construction projects. One critical advancement in this regard is the adoption of Building Information Modelling (BIM) technologies and workflows. BIM offers a platform for architects to collaborate with various stakeholders, from engineers to contractors, suppliers and even clients, in a more integrated manner. It facilitates the efficient exchange of information, reduces the risk of clashes, contradictions and misinterpretations and streamlines the entire construction process. Architects, when well-versed in BIM, can be pivotal players

Figure 10.1 Example of digital representation of MEP *(Mechanical Electrical and Plumbing)* supports in complex pharmaceutical plant (Krönert (2021) with permission of Springer Nature).

in ensuring that information flows seamlessly across the supply chain. A digital prototype of a building fosters coordination of design elements. It can enable design for prefabrication of components that can be shipped to site for final fitting or assembly. Prefabrication in construction involves the off-site manufacturing or assembly of building components in a controlled environment, away from the construction site. This method deviates from traditional on-site construction by producing elements like walls, floors and mechanical, electrical and plumbing racks in a factory. Standardisation plays a key role in prefabrication, as uniformity of sizes and specifications of components facilitates the manufacturing process and efficient assembly on the construction site. Shifting the on-site work to the controlled manufacturing environment results in reduction of waste, along with improved safety and quality control due to factory-controlled conditions. Furthermore, the off-site manufacturing process enables productivity and cost-efficiency work on the site, making prefabrication an attractive construction method (Stehn and Jimenez 2024). Nevertheless, the downsides of this approach have also been previously discussed. In the digital age, skilled architects may be empowered to have greater control over the building process. The development of computer technologies has provided architects with a range of design and construction tools that can manipulate the materials, assemblies and constraints to facilitate design for prefabrication. This digital transformation can allow architects to design increasingly complex shapes and structures, but generally, huge repetition of simple components is required. Three-dimensional models can now be translated into digital fabrication models, paving the way for greater precision and quality in construction. Prominent architects like Frank Gehry and Norman Foster have become known as virtual master builders. They design intricate projects that necessitate a close collaboration between architects and digital fabrication suppliers (Kolarevic 2005). This synergy allows these architects to push the boundaries of design and create buildings that were once considered nearly impossible to construct. Yet, they are in privileged positions, and this kind of attention to innovative forms and components is not generally available to most architects.

Thus, the use of BIM for prefabrication is not normally related to complex surfaces of buildings but more usually to greater efficiency through better information management and streamlining the design to the fabrication process. More accurate, reliable and ubiquitous data production with BIM can enable leaner production processes across the organisation or across the project and building supply chain. BIM is an enabler of process change towards the reduction of production cycle time, enhanced collaboration and teamwork and the reduction of waste by lowering time spent on detailing drawing production (Whyte and Hartmann 2017; Zhang et al. 2018). While few architects wield the same level of control over design and construction processes as megastars like Frank Gehry and Norman Foster, they have adapted to become essential contributors within digital workflows. The integration of technology, especially BIM, has the potential to revitalise their involvement in the entire project lifecycle. Architects now have the digital tools and expertise to encourage collaboration, provide data-driven insights and guide projects from conception to realisation. Through this evolution, architects are not just shaping spaces but also nurturing collaborative partnerships that drive innovation and create built environments that resonate with global trends. While the traditional notion of the master builder embodied in a single individual may no longer be feasible, architects play a crucial role in orchestrating the integration of diverse specialised disciplines within the field. The complex nature of some building projects necessitates a collaborative effort that draws on the expertise of various stakeholders, including clients, architects, engineers, contractors, facility managers, users and specialist designers. Although architects may oppose the designation of a BIM manager, it is important to acknowledge that this role extends beyond professional titles to encompass talent and experience. Handling data for complex projects requires a diverse skill set, and decisions based on BIM models may increasingly involve stakeholders beyond the architectural realm. The influence of owners, developers and financiers in assuming primary decision-making roles underscores the imperative for architects to adapt and stay attuned to evolving possibilities in this dynamic digital landscape (Deutsch 2011; Kostopoulos 2022), although it may have implications for the quality and appropriateness of the design.

In BIM projects, stakeholders collaborate around a BIM model, with each designer contributing parts to construct the virtual prototype. The BIM manager is crucial for integrating these diverse models created by different software packages, designers and contractors. The architect can play the role of BIM manager, given that the architect is typically the first stakeholder to be contracted by the client for a project under a traditional procurement system. A limitation here is that for many types of projects, other structures apply and the leading status of the architect is certainly not universal. Depending on procurement routes, contractors or consultants may also take the BIM manager role (Ahankoob et al. 2021). In arrangements, contractors, architects and engineers collectively contribute to project outcomes, sharing liabilities for structural integrity, safety and overall quality, adhering to industry standards. However, there are great unresolved difficulties over intellectual property rights, legal obligations and professional indemnity insurance. In this collaborative framework, BIM managers, responsible for managing the BIM model, play a pivotal role. Serving as a bridge between design and construction phases, they ensure the digital representation accurately reflects collaborative efforts. Effective coordination and adherence to established standards are essential to mitigate the risk of errors, contradictions and misinterpretations. Yet, BIM managers are increasingly computer professionals who know little about the design, construction and custodianship of the built environment.

Data-driven Design and the Architect's Role in Building Information Modelling

Over the past two decades, the adoption of computer-aided design and computer-aided manufacturing (CAD/CAM) technologies has had a significant impact on how automobiles, ships and aeroplanes are designed and produced. Paradoxically, the fragmented nature of the building industry hinders the implementation of business models that can facilitate the adoption of such integrated solutions (Schodek et al. 2005). The integration of design, developed in different systems, is achievable with BIM by increasing the collaboration of different disciplines in the design, construction and operation processes (Wong et al. 2010). In the contemporary digital landscape, the construction industry has witnessed the emergence of data-driven design as a fundamental principle (Langston and Zhang 2021; Philips and Azhar 2011). This paradigm shift positions virtual master builders as custodians of information, utilising BIM as a primary tool to leverage the power of data across the entire building lifecycle – from design and construction to operation. In building projects, several participants from different disciplines work on the project at the same time. Schodek et al. (2005) argue that there is a need to control concurrent design activities digitally, giving access or restraining access to certain parts of the project for certain teams or individuals. Kieran and Timberlake (2003) proposed that the problem-solving in the project design is possible through virtual visualisation and interaction with the models, adding that tools for information and communication are no longer physical but virtual. In these kinds of projects, BIM models have become a central data repository for the building data, but which data the model should contain, and at what level of detail, are aspects that are open to debate and will impact the overall BIM management process. The purpose of building information modelling varies depending on the perspective of those involved, and the kind of project or intervention in the built environment. In 2019, the International Organization for Standardization (ISO) directed its focus towards BIM to foster international collaboration in digital construction projects. As previously seen, the resulting ISO 19650 is a comprehensive international standard specifically designed for the management of information and documentation throughout the lifecycle of building assets. This standard represents a significant milestone in breaking down silos within the construction industry, particularly concerning data management in BIM projects. It simplifies the comprehension of responsibilities and boundaries within collaborative BIM workflows, serving as a bridge across various construction phases. From the viewpoint of the asset owner, establishing and upholding the purpose of the asset holds significant importance. Strategic business decisions, critical for guiding the overall direction of the endeavour, are central to this perspective. It involves the execution of comprehensive business plans, incorporating strategic asset portfolio reviews and life cycle cost analyses. Ensuring alignment with the project's purpose is pivotal for effective decision-making from the owner's standpoint. The user's perspective focuses on discerning authentic user requirements and ensuring that the asset solution possesses the required qualities and capacities. The project delivery or asset management perspective revolves around meticulous planning, resource mobilisation, effective coordination and development control. BIM execution plans, organisational charts and function definitions are crucial components from this management viewpoint. On a broader societal scale, considerations involve safeguarding community interests throughout the asset life cycle, with political decisions, area plans and regulatory processes like building permits and concessions. It is interesting that none of these considerations include the quality and appropriateness of the built environment or the concept of place.

Building information modelling stands as a pivotal technological innovation in the production of buildings, involving the creation and management of a comprehensive 3D model that encapsulates the physical aspects of a building (Azhar et al. 2012). This model also incorporates the functional characteristics of the construction components. Within the BIM model, both graphical and non-graphical information, including parameters linked to graphical data, coexist. The merging of graphical and textual information means that operation and maintenance manuals are a joint deliverable for the building facility managers and users. According to ISO 19650-1:2018, two essential components of BIM are the asset information model and the project information model. The asset information model supports strategic and day-to-day asset management processes established by the building owner who initiated the construction project, offering valuable insights at the project's outset. It encompasses diverse details such as equipment registers, cumulative maintenance costs, installation and maintenance dates, property ownership specifics and other pertinent information, all presented systematically. The project information model includes project geometry, equipment locations, performance requirements during project design, construction methods, scheduling, costing and details of installed systems, components and equipment, including maintenance requirements during project construction. At the core of BIM, there is data, ranging from geometric representations to detailed specifications, performance data and maintenance information. The efficient utilisation of this data is BIM's primary objective, aiming to foster collaboration, inform decision-making and enhance efficiency across the entire lifecycle of a building (ISO 19650-1:2018). Throughout the phases of project design, architects and engineers harness the power of BIM applications for activities such as option analysis, 3D modelling, energy modelling, scheduling and clash detection. This empowers decision-making while optimising design outcomes. During the pre-construction phase, the BIM model can support tasks such as quantity surveys, risk identification, safety planning and constructability analysis. During the construction phase, BIM allows monitoring of project progress through 4D phasing plans, construction sequencing and managing construction information effectively. In the post-construction and facility management phases, owners benefit from comprehensive BIM data, simplifying maintenance and assisting real estate portfolio management. Furthermore, as previously noted, the adoption of BIM opens opportunities for prefabrication, offering a more productive construction process by designing for off-site manufacturing. This process is about understanding how the design will be physically built and incorporating it throughout the design process, together with the manufacturer's specialist knowledge of fabrication while ensuring the architect retains control of the design intent. The constraints of modularity and repetition are essential to this interpretation of efficient design. Incorporating dimensional modules and using off-the-shelf component items can be challenging to the architect's creativity and the specifics of site, location and place. If these factors are carefully considered in the early stages of the design process, it can be determined if this is the correct route. Over the years, some product manufacturers have become involved with BIM and developed models of their products to share with architects. Once the concept design is concluded, the architect can send the models back to the manufacturer. These models may be linked to a manufacturer's database of products that contain a fully detailed model, including all information necessary for their digital fabrication. Two-way communication about the impact of requesting variations to standard products is essential to efficient delivery. However, not all products in construction are off-the-shelf components. Building projects are developed according to specific client's needs and built on a site with specific requirements. Therefore, the products must be designed specifically for the building, a concept called engineering to order. However, this may present difficulties for individual projects as components need to be supported by a production strategy so that the manufacturer can achieve volumes of production quality assurance and cost-effectiveness. For Kieran and Timberlake (2003), standardisation and customisation are real challenges for architecture and construction, but there may be a strategy to incorporate fabricated products while meeting the needs of users for choice, expression and individuality. There could be a hybrid in which fabricated products are functional and non-visual, whereas the appearance of the form and spaces are customised to each situation.

Not sure what has happened to the font size Digital fabrication is a process that utilises digital data to direct the fabrication of building components or assemblies (Whyte and Hartmann 2017). It consists of three major components: a CAD system for design, a CAM software to insert manufacturing information, and one or more CNC machines for fabrication (Schodek et al. 2005). Dunn (2012) argues that in architecture, the use of CAD/CAM systems has transformed not only the way architects represent their ideas but also how they generate their ideas. Moreover, the complexity of shapes and forms that can be achieved with the application of digital design for fabrication or even the use of algorithmic design would be inconceivable without the use of technology. On the other hand, understanding how the product or part of the building will be manufactured is essential to incorporate it into the design, as elements that will further facilitate digital fabrication (Kuo et al. 2001). Some architects and designers lack knowledge of CAD/CAM system functionalities. Architects can play an important role in digital design for fabrication, but they are not going to be working alone on this task. The complexity of some buildings requires several professionals to collaborate with their expertise. Each of them works on a different part of the model, for example, MEP designers, structural designers, manufacturers of the products and specialist subcontractors. Thus, the information management of BIM projects can be a complex task in ensuring satisfactory completion of the project. There are three main aspects to consider in data-driven design – the complexity of geometry, the complexity of the information included in the model parameters, and finally the costs associated with the delivery of the BIM model. The capabilities of a BIM model are enormous, but excessive and meaningless data can be costly and of little use. This can be illustrated by the example of a café in a Scottish airport. In this project, the architect was required to input the full set of construction-operations building information exchange data into the model. In this small project, the architect spent 500 hours designing the café and another 300 hours simply collecting and integrating the data for the model. Those hours came at a cost that was not well received by the users, who did not have a system in place to use this data during the operation of the café. However, they had to comply with the delivery of these data due to the requirements of the airport BIM execution plan. This is one of the concerns with systems, in that they may be inflexible and actually increase costs. Critical to the understanding of the data needed by a BIM model is the concept of the level of development. It serves as an indicator of the depth and accuracy of the information contained within the BIM model, measuring how comprehensively geometry and related data are considered at various stages of the project, spanning from initial design to construction and beyond. Bolpagni and Ciribini (2016) highlight various concepts related to the definition of data sets, including level of detail. Their findings indicate that there is no unequivocal approach to managing this issue, leading to the establishment of numerous definitions and acronyms in different countries, especially in the United States and the United Kingdom. However, a comparative analysis suggests the possibility of identifying similarities among these diverse approaches. The American Institute of Architects offers five distinct levels of development corresponding to various phases in the design and construction process (see Figure 10.2). These standards may facilitate effective collaboration among design teams, providing clarity regarding the status of design elements. It is crucial to note that the standards apply to individual building systems and components rather than the entire project. They serve as critical identifiers for specific design elements that have reached the construction documentation stage.

LOD 100 Conceptual	LOD 200 Approximate geometry	LOD 300 Precise geometry	LOD 400 Fabrication	LOD 500 As-built
The Model Element **may be** graphically represented in the Model with a **symbol or other generic representation,** but does not satisfy the requirements for LOD 200. Information related to the Model Element (i.e. cost per square metre, etc.) can be derived from other Model Elements.	The Model Element is graphically represented in the Model as a **generic** system, object or assembly with **approximate** quantities, size shape, location and orientation.	The Model Element is graphically represented in the Model as a **specific** system, object or assembly **accurate** in terms of quantity, size shape, location and orientation.	The Model Element is graphically represented in the Model as a **specific** system, object or assembly that is **accurate** in terms of quantity, size, shape, location, and orientation with **detailing, fabrication, assembly, and installation information.**	The Model Element **is** a **field verified** representation **accurate** in terms of size, shape, location, quantity and orientation.
	Non-graphic information may also be attached to the Model Element.	Non-graphic information may also be attached to the Model Element.	Non-graphic information may also be attached to the Model Element.	Non-graphic information may also be attached to the Model Element.

Figure 10.2 The table of levels of development by Jinhyeok Yang (eng BIM company LLC/https://engbim.com/what-is-lod/ / accessed March 20, 2024).

Architects can play a leading role in the realm of data management. Their responsibilities extend across the spectrum of BIM levels of development. These can encompass generating, overseeing and aligning data within BIM models to the appropriate level, ensuring the accuracy and consistency of the data. This role takes on added significance when architects assume the role of BIM managers, where they are tasked with validating data sourced from various project stakeholders. In this instance, they need to coordinate and align with project-specific goals and customer objectives, ensuring that the data serves the project's objectives effectively (Uhm et al. 2017). This comprehensive role positions architects as the custodians of data integrity within BIM projects, underscoring their vital function in maintaining data consistency and accuracy. The influence of BIM transcends the distinct phases of a construction project, enabling architects to streamline processes and make well-informed decisions. In the earliest stages of project programming, BIM empowers architects with insights into spatial considerations, space standards and land regulations. This facilitates the efficient analysis of project requirements and the identification of potential design challenges. Furthermore, BIM fosters data integration across the entire project life cycle, promoting enhanced communication and collaboration among project stakeholders (Azhar et al. 2012). In summary, it could be argued that architects have evolved into central figures in data management and BIM is the catalyst, enabling them to make informed decisions, achieve cost savings and efficiently develop their projects. This evolution could transform architects into information masters, empowering them to apply data-driven design principles that enhance collaboration and drive efficiency within architectural practice. The new significance of architects could be highlighted by their role in managing data within the context of BIM, aligning projects with data-driven design and realising innovative and sustainable architectural solutions (Philips and Azhar 2011). Nevertheless, there is some concern as to whether all architects view themselves in this way and whether such a mechanistic way of working is suitable for shaping the future of the built environment in different cultures around the world.

A Global Approach and Local Issues

The relentless pace of technological advancement has transformed the way that people live and work in IT-enabled countries. In construction, staying up-to-date with rapidly evolving technology is crucial for efficiency, safety and competitiveness (Demian and Walters 2014). However, this presents unique challenges for developing countries, where limited resources and access to technology can impede their ability to keep pace with developed nations. In this context, the impact of rapidly evolving technology on construction projects, and the challenges and opportunities it presents for architectural practices, will be explored in both developed and developing countries. The rapidity of change in the world of technology is a fact, and this is particularly evident in the constant evolution of software tools. As soon as one version is released, developers are already working on the next iteration, aiming to make it even more sophisticated and feature-rich. This rapid progression demands a significant level of adaptability and continuous investment for both individuals and organisations. Staying up-to-date with the latest software versions is essential for remaining competitive and efficient in today's digital landscape. This relentless cycle can present challenges, especially for those in developing countries who may struggle to keep pace due to limited education resources and access to technology. Developed nations often have the means to invest in innovative construction technologies, like software to implement BIM, drones and advanced construction machinery. The software packages offer powerful capabilities for design, collaboration and project management, which can enhance the efficiency and quality of architectural projects. However, they also come with significant costs, which can be a major blocker for architectural practices, particularly for small companies based in developing countries. Compatibility between earlier software versions and the most up-to-date versions is thus a critical issue for many businesses. One of the primary challenges faced by developing countries is a lack of access to the necessary resources and expertise. Acquiring and implementing advanced technologies requires significant investment in both infrastructure and training (Acheng et al. 2023). Without access to resources, projects in these nations may lag in terms of efficiency, safety and sustainability. Moreover, the construction sector in developing countries often relies heavily on manual labour, making it challenging to transition to more automated and tech-driven processes. Local knowledge remains a critical issue for developing countries, which are aware that international collaborators may not fully understand the reality of their situation. In some developing countries, projects are managed by traditional methods and regulations that may not readily accommodate advanced technological solutions. Integrating these technologies can require substantial changes to established norms, which can be met with resistance from industry stakeholders and government bodies. In remote or rural areas, reliable internet access may be scarce or non-existent, limiting the ability to leverage cloud-based software, remote monitoring systems and other digital tools that are essential for efficient and data-driven construction practices. This connectivity gap further widens the technology divide between developed and developing nations.

Although the transformation related to digital workflows is based on modern technologies, the implementation of BIM and the associated technologies requires substantial changes in organisational culture and process improvement. Mere acquisition of new software packages does not ensure that it will work, that lies in people embracing new ways of working facilitated by innovative technology. Thus, comprehensive organisational changes must be meticulously overseen throughout the BIM implementation process (Abdirad 2017). Change management strategies related to BIM should focus on ensuring that people engage with it, reducing resistance to change and planning the training. Thus, people skills grow as company capabilities increase (Arayici et al. 2011; Khosrowshahi and Arayici 2012). Change management, as recognised by leaders, may stem from external opportunities presenting themselves as a chance for internal organisations to change. However, merely acknowledging the need for change is insufficient to initiate it (Hayes 2014). According to Lewin (1952), reducing resistance to internal changes increases commitment and results in a more stable situation than when they are exerted by external pressures. Lewin's three-step model for organisational change includes unfreezing existing behaviour, acting and refreezing behaviour at a new level (Hayes 2014). Bringing this idea to the context of embracing digital workflows in architectural practices, more than investing in technology, leaders will invest in people to build up the teams' motivation to adopt the proposed changes in the working methods. In addition, a bottom-up approach, rather than a top-down approach can reduce the team's resistance to change (Arayici et al. 2011). Empowering individuals to embrace change rather than resist it is a strategy more likely to succeed than one of a top-down imposed solution. Notwithstanding, even this approach to first-world situations may have run its course. Concern is emerging that change as a permanent state is actually causing damage, and the benefits of continuity are being lost. Moreover, business paradigms are very much context based. While increased technology and change management may suit commercial markets with emphasis on realisable assets, they may not be appropriate for other spheres of activity. The difficulties experienced by developing countries have already been demonstrated. Yet, even in developed countries, there is a range of different situations. The concept of the project might not even be recognised. Minority cultures and local communities may not even engage with business speak. It is clear that one size does not fit all.

Conclusion

The evolution of architecture has undergone profound changes in roles and responsibilities, influenced by societal changes as well as technological advancements. Historically, some architects embodied the role of master builders, overseeing both visionary design and practical construction. However, the separation of design and construction processes over time led to a fragmentation of responsibilities and a less holistic approach to building projects. This shift towards specialisation created an adversarial atmosphere among stakeholders, emphasising the need for improved communication and coordination throughout the design and construction process. The evolution of the architectural profession has undergone various transformations, including more specialised roles and broader collaboration. Initially, a number of architects assumed dual roles as visionary designers and builders, but as the construction industry diversified, methods changed and responsibilities became more segregated. With the onset of large-scale projects, challenges in communication and coordination emerged. Although very visible in the built environment, these projects are only a part of design output and associated with the property industry or central governments. For these kinds of projects, BIM facilitates seamless data integration, reducing the risk of clashes and enhancing the overall efficiency of the construction process. Moreover, it enables architects to design for prefabrication, industrialising construction processes by producing building components offsite in controlled environments, thus improving safety and quality control.

ISO 19650 is a comprehensive guideline for managing information and documentation related to constructed assets. Adherence to ISO 19650 promotes international consistency and enhances efficiency in digital construction projects, marking a significant milestone in breaking down industry silos and promoting shared responsibilities within collaborative workflows. Architects can still play a central role in the realm of data management within BIM projects, serving as the custodians of data integrity throughout the project life cycle. This role gains added significance when architects take on the role of BIM managers, validating data sourced from project stakeholders and ensuring its accuracy and consistency. Beyond design, architects and BIM managers are tasked with coordinating data to align with project-specific objectives. This comprehensive role positions architects as information masters. However, the translation of BIM models into physical reality presents significant challenges. From a construction viewpoint, early collaboration may be important to avoid misunderstandings and inefficiencies, highlighting the need for integration between architects and contractors. Yet, the power balance in these arrangements needs to be carefully considered to ensure that design quality is not sacrificed for construction expediency. In the context of rapidly evolving technology, staying up-to-date is crucial for efficiency, safety and competitiveness in the construction industry. Alternatively, communities may be seeking more low-tech processes and products. Developing countries also face challenges due to limited resources and access to technology. While advanced digitalisation may be beneficial for large and complex projects, it needs to be recognised that throughout the world, that it is not the only work undertaken by architects and builders. Ironically, the fourth industrial revolution that spawned the massive increase in digitalisation may actually be reducing demand for these kinds of projects.

References

Abdirad, H. (2017). Metric-based BIM implementation assessment: a review of research and practice. *Architectural Engineering and Design Management,* 13(1), 52–78.

Acheng, P.O., Kibwami, N., Mukasa, T.J., Odongkara, B.B., Birungi, R., Semanda, J. and Manga, M. (2023). Building information modelling adoption in Uganda's construction industry. *International Journal of Construction Management,* 23(13), 2185–2208.

Ahankoob, A., Manley, K., Hon, C. and Drogemuller, R. (2021). The influence of building information modelling on the absorptive capacity of project-based organisations. *Architectural Engineering and Design Management*, 1–21. https://doi.org/10.1080/17452007.2021.1881879.

Arayici, Y., Coates, S. P., Koskela, L. J., Kagioglou, M., Usher, C. and O'Reilly, K. (2011). BIM implementation for architectural practices. *Structural Survey,* 29(1), 7–25. https://doi.org/10.1108/02630801111118377.

Asgari, S. and Noorzai, E. (2023). Improving the effectiveness and interaction between building information modeling and life cycle assessment. *Architectural Engineering and Design Management,* 19(1), 22–38. https://doi.org/10.1080/17452007.2021.1889956.

Azhar, S., Khalfan, M. and Maqsood, T. (2012). Building information modelling (BIM): now and beyond. *Australasian Journal of Construction Economics and Building,* 12(15). https://doi.org/10.5130/ajceb.v12i4.3032.

Bolpagni, M. and Ciribini, A. (2016). The information modeling and the progression of data-driven projects. *Proceedings of the CIB World Building Congress 2016,* Tampere, Finland (30 May–3 June 2016). 296–307. Tampere, FI: TUT – Tampere University of Technology.

Burr, K.L. (2011). What happened to the master builder?: implications for the built environment. *International Journal of the Constructed Environment,* 1(1), 1–14.

Burry, M. (2005). Between intuition and process: design and rapid prototyping. In B. Kolarevic (ed.) *Architecture in the Digital Age: Design and Manufacturing*. New York, London: Taylor & Francis. 147–162.

Demian, P. and Walters, D. (2014). The advantages of information management through building information modelling. *Construction Management and Economics,* 32(12), 1153–1165.

Deutsch, R. (2011). *BIM and Integrated Design: Strategies for Architectural Practice*. Hoboken, NJ: John Wiley & Sons.

Dunn, N. (2012). *Digital Fabrication in Architecture*. London: Laurence King.

Forgues, D., Koskela, L. and Lejeune, A. (2009). Information technology as boundary object for transformational learning. *Journal of Information Technology in Construction,* 14, 48–58.

Garber, R. (2014). *BIM Design: Realising the Creative Potential of Building Information Modelling*. Chichester: John Wiley & Sons.

Hayes, J.A. (2014). *The Theory and Practice of Change Management* (Fourth Edition). Basingstoke: Palgrave Macmillan.

ISO 19650-1:2018 (2018). *Organization and Digitization of Information About Buildings and Civil Engineering Works, Including Building Information Modelling (BIM) — Information Management Using Building Information Modelling — Part 1: Concepts and Principles*. Geneva, Switzerland: International Organization for Standardization. https://www.iso.org/standard/68078.html.

Khosrowshahi, F. and Arayici, Y. (2012). Roadmap for implementation of BIM in the UK construction industry. *Engineering Construction and Architectural Management,* 19(6), 610–635. https://doi.org/10.1108/09699981211277531.

Kieran, S. and Timberlake, J. (2003). Refabricating Architecture. New York, London: McGraw-Hill.

Kolarevic, B. (2005). Digital production. In: *Architecture in the Digital Age: Design and Manufacturing* (ed. B. Kolarevic), 29–54. New York, London: Taylor & Francis.

Kostopoulos, K. (2022). Collaborative practice-based learning methods in architectural design and building technology education in a cross-cultural, cross-geographical environment. *Journal of Architectural Engineering*, 28(1). https://doi.org/10.1061/(ASCE)AE.1943-5568.0000525.

Krönert, N. (2021). BIM bei Hilti. In A. Borrmann, M. König, C. Koch, and J. Beetz (eds.) *Building Information Modeling*. Wiesbaden: Springer Vieweg. 757–766. https://doi.org/10.1007/978-3-658-33361-4_39.

Kuo, T.C., Huang, S.H. and Zhang, H.C. (2001). Design for manufacture and design for 'X': concepts, applications, and perspectives. *Computers & Industrial Engineering*, 41, 241–260.

Langston, C. and Zhang, W. (2021). DfMA: towards an integrated strategy for a more productive and sustainable construction industry in Australia. *Sustainability*, 13(16). https://doi.org/10.3390/su13169219.

Lewin, K. (1952) Field theory in social science. New York: Harper and Row.

Philips, S. and Azhar, S. (2011). Role of BIM for facility management in academic institutions. *Proceedings of the 6th International Conference on Construction in the 21st Century (CITC-VI)*, Kuala Lumpur, Malaysia (July 5–7 2011), 950–957.

Schodek, D., Bechthold, M., Griggs, K., Kao, K., and Steinberg, M. (2005). *Digital Design and Manufacturing: CAD/CAM Applications in Architecture and Design*. Chichester: John Wiley & Sons.

Stehn, L. and Jimenez, A. (2024). Industrialized house building productivity growth. *Construction Innovation*, 24(7). https://doi.org/10.1108/CI-04-2022-0097.

Uhm, M., Lee, G. and Jeon, B. (2017). An analysis of BIM jobs and competencies based on the use of terms in the industry. *Automation in Construction*, 81, 67–98. https://doi.org/10.1016/j.autcon.2017.06.002.

Whyte, J.K. and Hartmann, T. (2017). How digitizing building information transforms the built environment. *Building Research and Information*, 45(6), 591–5. https://doi.org/10.1080/09613218.2017.1324726.

Wong, A.K.D., Wong, F.K.W., and Nadeem, A. (2010). Attributes of building information modelling implementations in various countries. *Architectural Engineering and Design Management*, 6(4), 288–302. https://doi.org/10.3763/aedm.2010.IDDS6.

Zhang, X., Azhar, S., Nadeem, A. and Khalfan, M. (2018). Using building information modelling to achieve lean principles by improving efficiency of work teams. *International Journal of Construction Management*, 18(4), 293–300. https://doi.org/10.1080/15623599.2017.1382083.

11

How Digital Projects are Changing Contracts

Erik A. Poirier, Ivanka Iordanova, and Gabriel Jobidon

University of Quebec, Montreal, Canada

Introduction

Contracts are essential in ensuring that parties entering into an agreement clearly understand their roles, obligations, the scope of their responsibilities and the compensation that will be provided (Hughes et al. 2015). Without a proper contract, the terms and conditions under which services or products are delivered remain undefined, which leaves room for misalignment between the expectations and requirements of owners and the incentives and obligations of service or product providers (Sande and Haugland 2015). The emergence of digitally enabled project delivery, with building information modelling at the helm, is transforming how projects are delivered and services provided. Indeed, digital projects are seen as redefining relationships between project stakeholders and the information flows between them (Succar and Poirier 2020). This redefinition is prompting changes to the legal and contractual landscape within projects, across organisational boundaries and throughout the broader industry. To address these changes, researchers and practitioners have been assessing the contractual aspects of digitally enabled project delivery to uncover the elements that must be addressed to ensure successful implementation and delivery. These elements include, among others, intellectual property, model and data ownership, liabilities, roles and responsibilities (Assaad et al. 2020; Winfield 2022). The way in which these elements have been identified and are being addressed mirrors a more evolutionary approach to digital implementation within the built asset industry, as opposed to the predominant revolutionary discourse. Considering that building information management we need to ensure consistency across the chapters is conceptualised as both an evolutionary and a radical innovation (e.g. Taylor and Bernstein 2009), there should be an appraisal as to what extent digital technology is changing how projects are procured and contracts developed. In parallel, as integrated approaches and models for project delivery are emerging to overcome fragmentation, BIM and other digital tools and processes are shining light on the opportunity for better integration of information across organisational boundaries and project lifecycle phases. While promising, this also poses significant challenges. There is an

Architectural Design and Management in the Digital Age: International Perspectives, First Edition.
Edited by Bob Giddings, Silvio Melhado, and Jenni Barrett.

assumption that digital projects are indeed changing contracts, and this notion should be reviewed. Certain elements within existing contractual frameworks are being modified, and there are calls for change, but in a broader sense, the characteristics of digital projects may require more fundamental and significant change. Current efforts aimed at addressing the integration of digital ways of working into existing standardised contracts offer a patchwork of solutions to discrete problems. To this effect, established and emerging standards and institutionalised practices will be discussed. The way in which contracts are changing will then be examined from two perspectives: in form and in function. Finally, novel approaches to project delivery, such as Integrated Project Delivery and Alliancing will be considered, and their influence, synergies and alignment with digital project delivery discussed from an asset lifecycle information management perspective. This includes how and for what purpose information is required, how it is delivered and exchanged, how and who manages it and how it is accessed and controlled.

Digital Projects and Contracts

The general premise is that digital projects are changing contracts. Considerations driving this change relate to questions as to how teams collaborate in digital environments and how information is exchanged between project stakeholders, who are parties to the contract. Indeed, contracts using traditional information sources and delivery mechanisms clearly delimit the precedence and scope of each contract document and distribute relevance of information. In cases of overlap or discrepancy, information precedence is prioritised according to provenance or substance, as laid out within the contract itself. Moreover, being delivered through traditional mechanisms and channels, namely documents and drawings, each of these information resources constitutes a piece of a larger proverbial contractual puzzle. This introduces significant potential for error and interpretation. With the push for integration of information through digitalisation, the idea is to eliminate redundancies and assure quality while making relevant information accessible to the people who require it. Contracts must evolve to enable these mechanisms to take place. In this regard, contracts are seen as heavily influencing the extent to which digital technology is being implemented at the project level. When considering factors affecting BIM implementation within project settings, legal and contractual issues consistently rank among the major barriers (Siebelink et al. 2021). Similarly, these issues are regularly included in BIM adoption and implementation frameworks, including normative and regulatory frameworks, which confirm their influence. As such, significant research has gone into identifying and defining the core elements that must be addressed to ensure that digital projects can be successful (Ahmed and Kassem 2018). In parallel, industry efforts around the world have led to the development of contract appendices and protocols, addressing some of the elements covered in the research, in order to structure how digital technology can be implemented in a project setting (Ragab and Marzouk 2021). Examples include the American Institute of Architects recent review of its BIM protocol (AIA 2022), the UK BIM protocol (UK BIM Framework 2021) and the Canadian BIM Contract Appendix (BuildingSMART Canada 2014), among many others. These resources aim to circumscribe the scope and use of digital technology, and BIM in particular, by addressing the core contractual elements

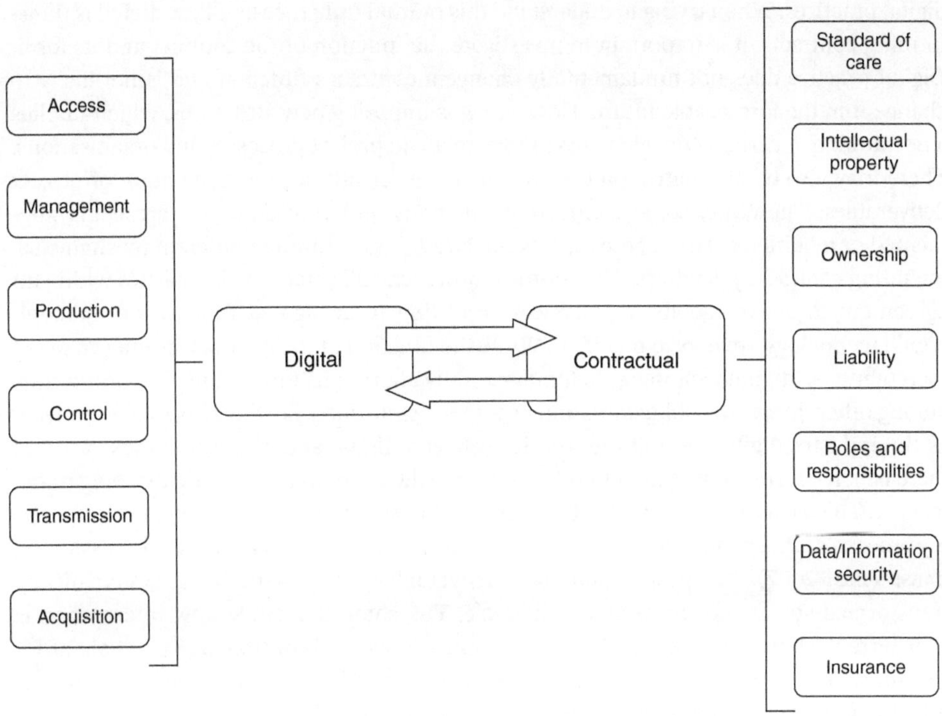

Figure 11.1 Interaction between digital projects and contractual considerations.

that have been identified as being critical. Yet, their adoption and use remain challenging (Winfield and Rock 2018). Figure 11.1 presents the interaction between digital projects and contractual considerations, detailing the defining elements from both conceptual domains. The arrows between the concepts illustrate their mutual influence. Whereas contractual considerations define the scope of implementation of digital technology, they also challenge current contractual elements and considerations. The characteristics of each domain have been developed and are discussed in the literature, e.g. Alwash et al. (2017). It can be observed that the contractual domain is in a reactionary stance with regard to digital practice. This means that contractual considerations and elements are continuously emerging. Indeed, digital delivery is a rapidly changing and transformative field that articulates several avenues of development. Moreover, the nature of digital projects, namely that they are posited as being more collaborative and integrative, has prompted the refinement of these core elements. From a contractual point of view, many elements merit further exploration.

How Digital Projects Are Transforming Contracts

As noted above, the contractual domain has been in a largely reactionary stance to the developments of digital technology. While this is a characteristic of the legal domain, due to the notion of jurisprudence or case law among others, this significant caveat will influence how and to what extent contracts are transformed and, conversely, can influence digital practices. When trying to understand this mutual influence and how digital is transforming contracts, it is important to investigate the function of the contract and its form. Digital practice does not fundamentally change a contract's function and is not likely to change it in the foreseeable future. However, it is imposing new structures, which implies a re-thinking of certain core elements as they relate to project processes and organisations. There may also be alternative outcomes, which can involve a re-consideration of project deliverables. Therefore, some contractual functions and considerations are being augmented or modified to suit. These aspects can be addressed through different mechanisms, including contract appendices. The contract, however, still serves to bind parties within an agreement to deliver goods and services according to an agreed framework, although digital technology appears to modify it. While the function of the contract is being changed incrementally, digital technology is fundamentally altering its form. Thus, it is prompting, among other things, the obligations and how they are fulfilled, i.e. the deliverables. Indeed, as the industry digitalises and the frontier between digital and physical worlds becomes increasingly blurred, outcomes of built asset lifecycle activities and phases are being transformed. The most salient example of digital transformation of the traditional project delivery processes is the handover of project and asset documents at the built asset commissioning phase (East 2007). Yet, project and asset deliverables are actually being systematically transformed across the entire project lifecycle. The changes include how information is transferred, where it resides and how it is commissioned and controlled (Kim et al. 2020). The morphing of the physical into the digital is supported by various mechanisms, such as integrated information environments, smart contracts and distributed ledger technology. BIM, smart contracts and ledger technology have received increasing attention in the past decade as they show promise in helping overcome some of the built asset industry shortcomings relating to the design, management and execution of contracts, as shown, for example, by Li et al. (2019). These technologies can help reduce complexity through automation and streamline transactions in a given setting (Mason 2017). However, as the notion of smart contracts and ledger technology in construction is relatively new, their implementation in industry is still in the very early stages.

Smart contracts are defined as a computerised transaction protocol that executes the terms of a contract. The general objectives of smart contracts are to satisfy common contractual conditions such as payment terms, reduce expectations and minimise the need for trusted intermediaries. Smart contracts function by automating micro-transactions independently from one another in accordance with a set of expected, yet clearly delimited, conditions and outcomes. The emergence of ledger technology enables these micro-transactions and supports transparency and trust in the process. The utility of BIM in the context of smart contracts becomes evident when evaluating automation and scoping of the works. This raises a number of different issues, especially in the legal field, which is subject to the particularities of each jurisdiction. A notable example are views on the acceptance of works. Most civil or common law jurisdictions stipulate that the acceptance of works should be done by the owner. These are usually considered as public order provisions, which means that they are mandatory and cannot be modified or restricted by parties entering an agreement. However, smart contracts often rely on decentralisation, which requires the verification of job performance by third parties and places such mechanisms in contravention of public order dispositions. Another example is the balance that needs to be struck between flexible contracts and the code executed by smart contracts, which may come into conflict since the input parameters and execution steps need to be specific and objective. While these traits may help carry out more administrative contractual burdens, such as ensuring funds are transferred from one account to another, they may fall flat in more complex and relational functions of a contract such as negotiation of a change order or unanimous decision-making by a steering committee regarding the settlement of a dispute. This means that contracts should ensure an interplay with traditional text agreements and softer subjective legal obligations while also providing space for smarter operations. Therefore, digital projects transform contracts in two specific ways. On one hand, the elements and considerations found within contracts are slowly evolving, in a largely reactionary fashion. On the other hand, the form that contracts take is set to be transformed significantly through digitalisation. In both cases, however, the context in which this happens will dictate the extent of this transformation and more importantly its potential. Indeed, the way in which projects are procured determines the contract terms, which in turn influence how digital projects are run by dictating how obligations are outlined, how they are to be met and by whom. Research has shown that the better the alignment between a project's structure imposed through the procurement approach, a project's requirements and expectations, the project team members' capabilities and intentions and the incentives, the better the project outcomes (Poirier et al. 2017).

Pathways Towards Integrated Digital Practice

Digital practice is purported to achieve its full potential through integrated structures (Fischer et al. 2017). These structures are, as discussed, enacted through contracts. Current contractual practices are largely seen as fragmenting roles and responsibilities and clearly delineate between the provision of different services and their outcomes. This contractual fragmentation has been shown to impact the effectiveness of these services and the quality of outcomes. While there are many reasons and motivations that lead to the choice of a specific procurement approach and a subsequent contractual set-up, consensus is emerging around the fact that better integration of project stakeholders will lead to more satisfactory project outcomes (El Asmar et al. 2013). When it comes to digital practice, the mitigation of contractual barriers through better integration helps to streamline the flow of information and the use of informational deliverables. In parallel, digital practice, as manifested through BIM implementation, can positively impact collaboration, if the appropriate structures are in place (Poirier et al. 2017). As such, digital and collaborative practices have been seen to produce more satisfactory outcomes through high levels of project team integration (Fischer et al. 2017). Currently, integrated project delivery, partnering or alliances, as developed in standard contracts such as NEC4 in the United Kingdom, Consensus Docs series 300 in the United States and the CCDC 30 in Canada, combined with standardised information management practices through the ISO 19650 series of standards, offer a pathway to achieve these outcomes. Digital practice and its operationalisation through BIM are currently being framed through the ISO 19650 series of standards. These aim to formalise information production and management methods and processes, which span the entire lifecycle and cover assets or a portfolio of assets, of all types and sizes. The series of standards is mainly articulated around five key information management principles – information requirements as determined through specified purposes delivered through use cases, information procurement, information delivery planning, information delivery and information acceptance. Underlying these principles are specific information management functions and responsibilities. In addition, there are digital information ecosystems or common data environments that support these digitally enabled asset lifecycle information management practices. The standards are implemented through a series of activities and resources that are undertaken by different stakeholders. The core contractual document within the ISO 19650 framework is the BIM protocol, which sets out the key contractual elements discussed above (UK BIM Framework 2021). When considering the broader information and consequently data management practices, many general elements are included in the ISO 19650 series of standards. Yet, being a broader practice applied across domains, data and information management have their own set of characteristics (KPMG and Atkins 2021). Integrated project delivery has been developed over the past two decades and is gaining momentum around the globe as a highly performing procurement approach, meaning that it consistently drives cost savings for project stakeholders. Its core principles have been discussed and explored in existing research (e.g. Fischer et al. 2017). They are intrinsically tied to lean philosophy in production and put efficiency and transparency front and centre. Integrated project delivery aims to remove barriers to the process and enables the seamless flow of ideas and information within the team. Figure 11.2 illustrates the ties between the principles that characterise all three concepts. Ultimately, enabling digital practice and ensuring the best possible outcomes within a project context could be achieved through application and development of specific integrated project delivery (IPD) attributes. While most of the issues in a certain domain will be related to one or more aspects in another domain, there is one key IPD principle that is the most significant, and which will be impacted by or will impact the implementation of all other objectives. It is the aim of mutual respect and trust.

One of the first principles in enabling integrated digital delivery is ensuring clear, coherent and consistent data requirements, as operationalised through information uses and established to meet specific purposes. This process should involve key stakeholders early in the project and evolve throughout the project as opportunities arise. Moreover, identification of information purposes, uses and requirements is tied to project goals and expected outcomes, which in turn are related to benefits and rewards, as incentives for project stakeholders. ISO 19650 indicates that the appointing party should formulate a series of information requirements and formalise them within the tender documents as exchange information requirements. They are developed from an appointing party's organisational information requirements, which are subsequently articulated as project information requirements (ISO 2018). Appointed parties can formulate and add to their own requirements. They must be clearly articulated throughout the project lifecycle and formally received, addressed and accounted for by all relevant parties, thus confirming the importance of appropriate contractual mechanisms. As the information requirements form the basis for how digital practice will unfold for specified purposes, delivered through information uses, their identification and formalisation are critical steps. Integrated project delivery will technically allow a team to continuously assess and improve the information requirements throughout the project, especially in a context where most built asset owners

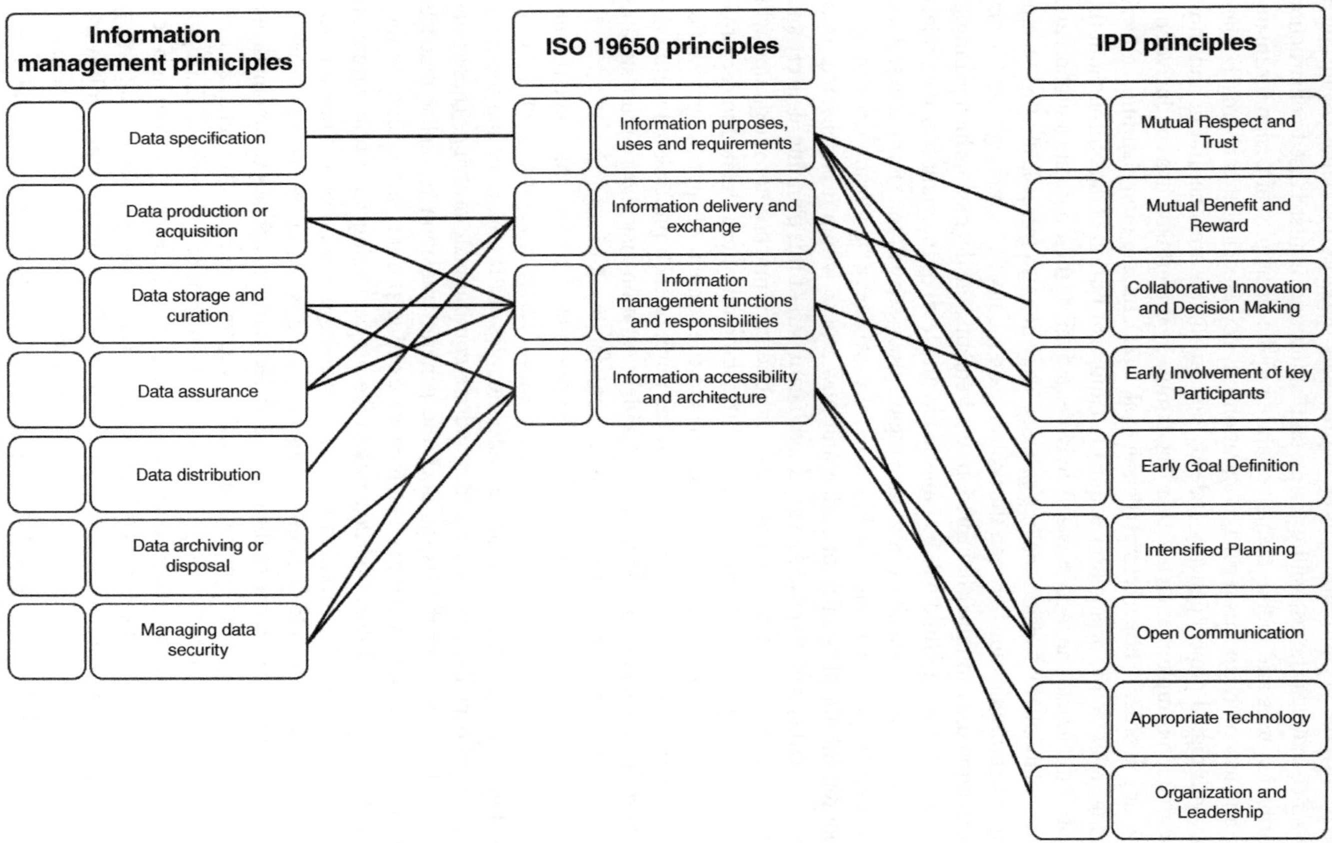

Figure 11.2 Integrated digital delivery: interaction between the principles of information management, ISO 19650 and IPD (adapted from Poirier et al. (2021) with permission of Emerald Publishing Limited).

find it challenging to formulate their information requirements (Cavka et al. 2017). Particular examples of the synergies between digital project delivery and integration, and more specifically, the need to rethink how contracts are developed abound. For instance, digital practice extends the capability for prefabrication and industrialised construction with information uses and requirements pertaining to the principles of design for manufacturing and assembly. In this regard, integration is a key aspect of this approach when combined with digital project delivery. Indeed, these attributes are a natural fit and show great potential to be implemented together. Yet, it must be undertaken in a conducive contractual setting in order to be successful (Poirier et al. 2021). Information planning, delivery and exchange relate to the production, acquisition and distribution of data throughout the lifecycle of a built asset. The processes and activities as well as their outcomes support and inform the collaborative decision-making process. They are also intrinsically tied to supporting open communication between the project team members. Stemming from the targeted information uses and their information requirements, the degree to which planning, delivery and exchange are fluid will be impacted by the level of integration of the project team. Moreover, as discussed above, although digital processes will transform deliverables, so will integration. As such, integrated digital delivery provides a significant opportunity to completely rethink the flow of information and how and to whom it is delivered. Data assurance and quality control processes are a significant part of this planning, delivery and exchange process. It is important to note that neither digital practice nor integration will fundamentally change professional liability and the standard of care related to each role (Manderson et al. 2017). Therefore, the assurance of the content of specific information deliverables, as they change states, will remain the remit of each professional within the integrated project delivery team. On the other hand, integration may allow a more streamlined assurance process with other factors such as buildability, for instance.

Integrated digital practice and project delivery impose new functions, roles and responsibilities. These will relate to the production, acquisition, curation, assurance and security of data within digital ecosystems. As an example, the roles of BIM manager and coordinator have emerged over the past two decades and are continuously evolving. These roles require new skills and capabilities, as does the general transition to digital practice. Moreover, these roles are sometimes assumed by existing project participants, or they sometimes require the additional participants to the project setting, such as information management consultants. These will impact on the provision of services and the numbers and types of contracts. The role of external consultants can be extremely useful in various contexts, for example, when some of the existing project participants lack digital maturity, or when the BIM capacity of a team is temporarily exceeded. They can also act as an integrator of the digital content for the project. In order to ensure the effectiveness of the digital delivery of a project in such circumstances, the consultants should be involved as a real party in the contract, with clear responsibilities and rewards. Another beneficial strategy for raising BIM maturity is for external consultants to train the in-house teams while working on a project with them and to make them autonomous for the next projects. These new roles and responsibilities, and the capabilities that support them, are increasingly being redefined and required in tender documents and contracts (Jobidon et al. 2021). Seen through the lens of integration, these new functions, roles and responsibilities speak to the necessity of early involvement of key participants as well as the organisation of the team and leadership within it. To ensure all people's capabilities and intentions, procurement processes focusing on effectiveness are necessary, whether through prequalification of suppliers, collaboration during the tendering processes and/or the use of quality selection criteria (Jobidon et al. 2021). Many procurement frameworks still involve the use of design-bid-build delivery methods and the use of the lowest compliant bidder approach. This hinders a team's ability to implement digital practices since important parties join too late in the process to add value to the project. However, procurement models favouring the integration of team members in earlier stages of the design can benefit digital projects. Early contributions of key suppliers help develop functional specifications and facilitate information management, communication and collaboration. These types of procurement models vary but usually consist of two-stage processes, which ensure early involvement of the contractor, main suppliers and manufacturers. The first stage, the request for qualification, ensures a more objective form of quality assessment through the examination of capacity, capability and experience. If a consortium has a sufficient score, it proceeds to the second stage, the request for proposals. The use of qualification notably helps to minimise the risk of selecting an unsuitable partner for a project by shortlisting participants and proposals at the same time. Different frameworks and tools enable owners to assess BIM capability, such as through BIM Quickscan. Succar et al. have developed five metrics to measure BIM performance, which could at least be partially used to assess firms' capabilities. The United Kingdom has also adopted a standardised questionnaire, the PAS 91:2013, which harmonises different specifications and helps public bodies reduce waste in their procurement processes. Questions 04–Q2 to 04–Q4 pertain to (i) the capability of working with a project using a common data environment; (ii) having documented policies, systems and procedures to achieve level 2 BIM and (iii) capability of developing, delivering and working to a BIM execution plan. This reduces legal uncertainties and allows to ensure interoperability of software and models, process harmonisation and staff training.

Most of the procedures already discussed focus on more objective criteria instead of elements such as behaviour. These elements are usually evaluated during the second stage of the two-stage procurement process, i.e. the request for proposals. This stage enables parties to collaborate through commercially confidential meetings, which can facilitate team integration and ensure discussions regarding digital requirements (Jobidon et al. 2021). More formal interaction processes complement those meetings, namely requests for information and optimisation measures. Interviews with senior management members or key participants can also be held and scored during the procurement process. A similar process is adopted for collaboration, where teams can be evaluated through scenarios or by using psychometric tests. These measures ensure a more comprehensive evaluation of tenderers while still ensuring fairness, as the evaluation process is itemised in the contractual documents and complements the evaluation of teams' digital capabilities. Finally, procurement documents should clarify which model serves as a reference for the realisation phase of a project. There can be some uncertainty about whether it is the model developed by the design and compliance team or the one developed by tenderers during the process. If the latter is deemed acceptable following the compliance evaluation of the proposal, it should serve as the basis for future works to ensure continuity and clarity (Jobidon et al. 2021). As far as intellectual property is concerned, architects and engineers can usually claim copyright ownership on drawings since they represent the fixation of an original idea as defined, for example, by Canada's Copyright Act. This act also provides for joint authorship in case of collaborative or collective works. Some searchers believe that not providing for joint authorship could hinder collaboration, especially with BIM level 3. Nevertheless, since BIM aims to ease facility management, models, information and data, it also holds significant value for the building owner. Thus, some contracts provide exclusive possession of models for the owner, and project data are transferred to assist good maintenance and trouble-free operation following completion (Jobidon et al. 2021). In exchange, owners provide designers and contractors a licence to use the intellectual property. This situation is becoming quite a usual approach, since most BIM manuals provide for these terms. However, other agreements ensure designers retain intellectual property. Parties should ensure that these topics are initially addressed in procurement documents and are discussed throughout the procurement process. The advantage of the owner maintaining the intellectual property rights is that it avoids split responsibility and potential difficulties with different indemnity insurance agreements. Where there is liability arising from the model and the information it contains, the origin is risk allocation being blurred by the intertwining of responsibilities. This is exemplified by collaborative decision-making, and parties need to address these issues as early in the process as possible. Recent case law has emerged regarding collaborative environments and digital practice, notably pertaining to the integrity of shared information, standard of care, professional negligence and admissibility of models as evidence. It is in this context of integrated and collaborative approaches, integrated project delivery and alliance that they may be useful to provide liability limitations or waivers to parties (Jobidon et al. 2021), although the legal situation is not yet fully resolved. Within the project setting itself, increasing reliance on digital technology to support project delivery activities calls upon specialised roles and functions to enable it to function. As integrated project delivery imposes its own structure and organisation of work through the senior management teams, project management teams and project implementation teams, where the division of labour is based on set design principles, digital deliverables will be structured accordingly. Ensuring the right members are found within each team and that they can all operate and contribute to the design and decision-making processes becomes a priority. Moreover, ensuring the quality of the data and having the appropriate capabilities to create it within the team, will ensure trust in the information as the project progresses. Access to information through integrated information ecosystems, or common data environments involves information storage and curation, archival or disposal and concerns data security. The notion of information ecosystems and data environments is evolving as the technology progresses. A common data environment is currently defined as an agreed source of information for any given project or asset for collecting, managing and disseminating each information container through a managed process (ISO 2018). It can be seen as both a workflow and a digital solution. The workflow described in ISO 19650 defines the positioning of specific information, as work-in-progress, shared, published or archived, which ensures that members of a project team use the information in the way that it was intended to be used. According to the standard ISO (2018: 25): *The advantages of adopting data environment solution and workflow include:*

- *Responsibility for the information within each information container remains with the organisations that produced it, and although it is shared and reused, only that organisation is allowed to change the contents;*

- *Shared information containers reduce the time and cost in producing co-ordinated information; and*
- *A full audit trail of information production is available for use during and after each project delivery and asset management activity.*

If implemented correctly, these ecosystems support open communication and transparency, improve data quality and consistency, clearly allow to identify ownership of data, enable easy exchange of models and data and provide clear communication of its status. This should allow for greater reliability of data, better spatial coordination and assist the generation of documents required by different team members (Mordue 2018). However, many barriers to data environments persist. For instance, setting up, running and managing one must be done by a party to the contract. Owners are typically identified as the ones with responsibility for setting them up (UK BIM Framework 2021), as they are seen to benefit the most, but they may not have the capacity or capabilities to do so. In addition, work is still largely performed through file-based exchange. Yet, it is tending towards information exchange at the object level due to traceability, control and assurance capabilities being addressed and developed within commercial tools. Conventions around information container naming, structure, etc. are being developed and diffused to ensure common ground and consistency. These must be agreed to, and formalised, by the project team. Even if owners become more aware of the benefits of digital delivery for the operations and maintenance of their facilities, they rarely have clear requirements to include in the contract. In addition, some of the widely used platforms are quite expensive, and the interoperability between platforms is rarely assured. As deployment and use of these environments continue to grow, their impact and implications from a legal perspective will continue to expand. For instance, ownership, curation and maintenance of these platforms, access to and responsibility for data found within, etc. are considerations that arise when implementing a platform across organisational boundaries in a project setting. To remove any ambiguity around such matters, contracts between parties must include provisions for addressing them. In an integrated setting, the project team takes on the liability for such an environment collectively. Indeed, participating organisations and signatories to the multi-party agreement generally waive liabilities, including those related to information and data exchange and use. For instance, liability limits are provided in the CIC BIM Protocol, which states that there is no warranty for the integrity of electronic data transmission or for corruption or alteration occurring after transmission (Mosey et al. 2016). Nevertheless, some exceptions to liability waivers are usually explicitly provided in standardised contracts and pertain to digital practice. Therefore, claims pertaining to violations or alleged violations of intellectual property rights, including infringements or alleged infringements of any patent or right copyright, or infringement or alleged infringement of any trademark or licence, are not covered by waivers. It is the same situation in the loss of work resulting from faulty design, construction or production of work or defects in the ground. Notwithstanding, liability waivers or limitations benefit digital practice and collaborative environments. The usefulness of BIM for traceability could lead to more identification of individual parties at fault. This kind of attitude has hindered the construction industry for long enough. Establishing a no-blame culture, open book approach and transparent environment benefits teams and project performance.

Conclusion

The discussion on the impact of digital practice and digital project delivery on contracts has been set out in three specific ways – the reciprocal relationship between digital technology and contracts, how digital technology is transforming contracts and the synergies between the principles of information management and integrated project delivery. The overarching objective is to establish pathways to streamline digital practice and digital project delivery by identifying the areas that are being reconsidered from both a contractual and a practical point of view. As such, past research characterising contractual aspects and elements, as well as characteristics of digital practice, has been identified and discussed. What emerges is that digitalisation of practice will not fundamentally change the function of a contract in the foreseeable future. Instead, its articulation and the provisions found in it are evolving to account for digital ways of working. Therefore, ownership of data and the ecosystem in which it is hosted, liabilities concerning access to and sharing of data, rights, roles and responsibilities, among others, are all being called into question, developed, and clarified. These changes are being assimilated into current standardised contract templates through specific appendices, exhibits and protocols. The move towards integrated project teams eases some of the questions around liabilities, roles and responsibilities, in that they are shared across the project team with the most appropriate members undertaking specific tasks. Moreover, the notion of shared risk and reward, and open book accounting can help in attributing modelling and coordination duties to the most capable and impacted team members. This move does not eliminate professional liabilities and obligations, nor does the transition to digital project delivery. Alternatively, the form of contract and its outcomes are set to change through technologies and solutions such as smart contracts. As a nascent, yet rapidly evolving, field of research, the dematerialisation of contracts and the deliverables tied to them require further investigation. Combined with integrated project delivery mechanisms, mainly related to open book accounting and full supply chain transparency, digitalisation shows much promise in streamlining delivery processes and making this approach even more relevant. However, many barriers remain to transition to integrated digital delivery, with socio-technical and cultural barriers chiefly among them. Thus, while digital capabilities extend the realm of the possible, as has been true throughout the past two decades of digital transformation within the built asset industry, people and processes will remain at the centre. Questions remain as to how to alleviate, facilitate and render more intuitive digital workflows and practices in an integrated project setting. Previous research and practice have demonstrated the vast potential for both improving outcomes and performance of the industry, yet there still remains a significant amount of work to ensure that the entire system is optimised, taking into consideration the various elements that characterise these emerging approaches. Current efforts towards a digital transformation offer a patchwork of solutions that address specific and limited parts of the whole. It may be that the present situation is where digital technology should be located. The variety found in the development of the built environment may not be suited to one method fitting all. There are parts of practice where it is clearly advantageous, but equally, there may be others where alternative routes are more appropriate.

References

Ahmed, A.L. and Kassem, M. (2018). A unified BIM adoption taxonomy: conceptual development, empirical validation and application. *Automation in Construction*, *96*, 103–127. https://doi.org/10.1016/j.autcon.2018.08.017.

AIA. (2022). Introducing AIA contract documents' 2022 BIM documents. *American Institute of Architects*. https://learn.aiacontracts.com/articles/6523765-introducing-aia-contract-documents-2022-bim-documents/.

Alwash, A., Love, P.E.D. and Olatunji, O. (2017). Impact and remedy of legal uncertainties in building information modeling. *Journal of Legal Affairs and Dispute Resolution in Engineering and Construction*, 9(3), 04517005. https://doi.org/10.1061/(ASCE)LA.1943-4170.0000219.

Assaad, R., El-adaway, I.H., El Hakea, A.H., Parker, M.J., Henderson, T.I., Salvo, C.R., and Ahmed, M.O. (2020). Contractual perspective for BIM utilization in US construction projects. *Journal of Construction Engineering and Management*, 146(12), 04020128. https://doi.org/10.1061/(ASCE)CO.1943-7862.0001927.

BuildingSMART Canada. (2014). *IBC 100-2014: BIM Contract Appendix* [Mandate-Contract or Agreement]. IBC-Institute for BIM in Canada. https://buildingsmart.gilmoreglobal.com/en/product/2422758f-c8a5-442d-a218-b512e2987515.

Cavka, H.B., Staub-French, S. and Poirier, E.A. (2017). Developing owner information requirements for BIM-enabled project delivery and asset management. *Automation in Construction*, 83, 169–183. https://doi.org/10.1016/j.autcon.2017.08.006.

East, E.W. (2007). *Construction Operations Building Information Exchange (COBIE): Requirements Definition and Pilot Implementation Standard* [Report]. Construction Engineering Research Laboratory (U.S.). https://erdc-library.erdc.dren.mil/jspui/handle/11681/19704.

El Asmar, M., Hanna, A.S. and Loh, W.-Y. (2013). Quantifying performance for the integrated project delivery system as compared to established delivery systems. *Journal of Construction Engineering and Management*, 139(11), 04013012. https://doi.org/10.1061/(ASCE)CO.1943-7862.0000744.

Fischer, M., Ashcraft, H.W., Reed, D. and Khanzode, A. (2017). *Integrating Project Delivery*. John Wiley & Sons. https://www.wiley.com/en-ca/Integrating+Project+Delivery+-p-9780470587355.

Hughes, W., Champion, R. and Murdoch, J. (2015). *Construction Contracts Law and Management* (Fifth Edition). Routledge. https://doi.org/10.4324/9781315695211.

ISO. (2018). *ISO 19650-1:2018 Organization of Information About Construction Works—Information Management Using Building Information Modelling—Part 1: Concepts and Principles*. https://www.iso.org/standard/68078.html.

Jobidon, G., Lemieux, P. and Beauregard, R. (2021). Building information modeling in Quebec's procurement for public infrastructure: a case for integrated project delivery. *Laws*, 10(2), Article 2. https://doi.org/10.3390/laws10020043.

Kim, S., Poirier, E.A. and Staub-French, S. (2020). Information commissioning: bridging the gap between digital and physical built assets. *Journal of Facilities Management*, 18(3), 231–245. https://doi.org/10.1108/JFM-04-2020-0024.

KPMG & Atkins. (2021). *The Value of Information Management in the Construction and Infrastructure Sector* (p. 104) [Text]. https://www.cdbb.cam.ac.uk/news/value-information-management-construction-and-infrastructure-sector.

Li, J., Greenwood, D. and Kassem, M. (2019). Blockchain in the built environment and construction industry: a systematic review, conceptual models and practical use cases. *Automation in Construction*, 102, 288–307. https://doi.org/10.1016/j.autcon.2019.02.005.

Manderson, A.D., Jefferies, M.C. and Brewer, G.J. (2017). An analysis of the integration of building information modelling (BIM) in standard construction contracts. In P. Wu, H. Li and X. Wang (eds.). *Integrated Building Information Modelling*. Sharaj, UAE: Bentham Books, 82–101.

Mason, J. (2017). Intelligent contracts and the construction industry. *Journal of Legal Affairs and Dispute Resolution in Engineering and Construction*, 9(3), 04517012. https://doi.org/10.1061/(ASCE)LA.1943-4170.0000233.

Mordue, S. (2018). *Implementation of a Common Data Environment: The Benefits, Challenges & Considerations*. The BIM Delivery Group for Scotland, Scottish Future Trust (SFT). https://www.scottishfuturestrust.org.uk/storage/uploads/cdeimplementaionresearchaug18.pdf.

Poirier, E.A., Forgues, D. and Staub-French, S. (2017). Understanding the impact of BIM on collaboration: a Canadian case study. *Building Research and Information*, 45(6), 681–695. https://doi.org/10.1080/09613218.2017.1324724.

Poirier, E., Staub-French, S., Pilon, A., Fallahi, A., Teshnizi, Z., Tannert, T. and Froese, T. (2021). Design process innovation on brock commons tallwood house. *Construction Innovation*. https://doi.org/10.1108/CI-11-2019-0116.

Ragab, M.A. and Marzouk, M. (2021). BIM adoption in construction contracts: content analysis approach. *Journal of Construction Engineering and Management*, 147(8), 04021094. https://doi.org/10.1061/(ASCE)CO.1943-7862.0002123.

Sande, J.B. and Haugland, S.A. (2015). Strategic performance effects of misaligned formal contracting: the mediating role of relational contracting. *International Journal of Research in Marketing*, 32(2), 187–194. https://doi.org/10.1016/j.ijresmar.2015.02.002.

Siebelink, S., Voordijk, H., Endedijk, M. and Adriaanse, A. (2021). Understanding barriers to BIM implementation: their impact across organizational levels in relation to BIM maturity. *Frontiers of Engineering Management*, 8(2), 236–257. https://doi.org/10.1007/s42524-019-0088-2.

Succar, B. and Poirier, E. (2020). Lifecycle information transformation and exchange for delivering and managing digital and physical assets. *Automation in Construction*, 112, 103090. https://doi.org/10.1016/j.autcon.2020.103090.

Taylor, J.E. and Bernstein, P.G. (2009). Paradigm trajectories of building information modeling practice in project networks. *Journal of Management in Engineering*, *25*(2), 69–76. https://doi.org/10.1061/(ASCE)0742-597X(2009)25:2(69).

UK BIM Framework. (2021). *Information Protocol to support BS EN ISO 19650-2 the Delivery Phase of Assets_Ed3* [Protocol-Manual].

Winfield, M. (2022). Legal implications of digitization in the construction industry. In D.M. Bolpagni, R. Gavina and D. Ribeiro (eds.) *Industry 4.0 for the Built Environment* (Vol. 20. Springer International Publishing. 391–411). https://doi.org/10.1007/978-3-030-82430-3_17.

Winfield, M. and Rock, S. (2018). *The Winfield Rock Report: Overcoming the Legal and Contractual barriers of BIM*. https://wearenima.im/wp-content/uploads/2022/09/The-Winfield-Rock-Report.pdf.

12

Roles and Responsibilities in the Digital Design Team

Mary Hardie

Western Sydney University, Australia

Context

The traditional analogue era of building project procurement involved clearly defined roles for consultants and contract-delimited areas of responsibility for all team participants. Many people were comfortable with this arrangement and found change difficult. However, the digital world of current-day design teams is quite different, partly because technology has enabled both synchronous and asynchronous interactions between team members across the globe. Of course, there are jurisdictional and contractual aspects to procurement that impact how project roles play out in different places. Generally, digital technology can foster integrated team processes, but it can also call into question leadership protocols and decision-making systems from the past. Design responsibilities are becoming more network-based and less hierarchical or driven by professional status. Coordination has always been critical to the success of the design process, but contractual forms and individual relationships have not always kept up with this more distributed network of decision-makers. A landscape of varying team structures and sometimes contested responsibilities has resulted from digitalisation and this is yet to be fully resolved. In other words, the process of delivering a digitally designed project has advanced unevenly in terms of full team integration. In some parts of the world, there has been a predominance of design and construct procurement in commercial projects for some time and that emphasis informs much of the ideas in this chapter (Doloi 2008). In this situation, the design team's obligation to the client is transferred to the contractor, and the responsibility for detailed design and project delivery is similarly moved to the contractor. This process is not without its potential for misunderstanding and disputation concerning the original design intent (Nolan 2018) and there are concerns about design quality in this form of procurement. In particular, relationship to place has been criticised. Nevertheless, commercial clients have strongly favoured the assurance that such package deal arrangements can give them on cost and project duration. It is possible that more recent developments, such as blockchain as a distributed ledger, will change the project balance again by automating tasks such as

Architectural Design and Management in the Digital Age: International Perspectives, First Edition.
Edited by Bob Giddings, Silvio Melhado, and Jenni Barrett.

sharing risk and reward while enhancing intra-project information flow in real time (Elghaish et al. 2022). As a result, it is possible that aspects of design intent may be more closely tracked through the design and delivery process by the original design team. It is acknowledged that the context described here is not universal and is open to reinterpretation across jurisdictions, local practices and global circumstances.

Collaboration across the project team, from design to delivery, has been demonstrated to have many benefits for quality project delivery. Early contractor involvement can potentially address a knowledge gap around delivering newer design aspirations for a high level of sustainability (Ferme et al. 2018). It has also been suggested that construction safety risks can be more successfully addressed by collaboration in the early design phase (Zou et al. 2008). Such integrated project design and delivery has been promoted as an ideal for some time in different national contexts (Shen and Walker 2001). Replacing a previous tendency towards adversarial relationships between designers and contractors with enhanced collaboration is clearly desirable, but the actual structure of this collaboration is, as yet, unclear. Several different models have been tried. Partnering and alliancing on projects has been suggested as part of the solution to break down the adversarial nature of the traditional project delivery pattern (Manley et al. 2009). Alongside the move towards greater collaboration, the rapid expansion of digital tools and the quantities of data that can now be manipulated may result in significant breakthroughs in managing project outcomes for the benefit of the parties actively involved and the community more generally. Recent research has indicated that significant gains in both productivity and sustainability can be made through the greater use of strategies such as offsite manufacture and Design for Manufacture and Assembly or design for manufacturing and assembly (DfMA) (Langston and Zhang 2021). At the same time, new project roles such as BIM managers and coordinators may have an impact on facilitating shared understanding across the whole delivery team with a best-for-project attitude. High-performing construction organisations are concentrating on improving their management practices to take advantage of the potential gains (Ershadi et al. 2021). Especially on sophisticated projects, the design and delivery of buildings is becoming a functional combination of data and expertise deployed to better predict the real outcome of the intended project desire. The challenge is to achieve the best mix of people and technology in order to address the seemingly insoluble environmental and social problems that are faced in the built environment. This focus on real outcomes should improve public perceptions of the industry in terms of trust and reliability.

Evolving Roles and Team Balance

Increasingly complex projects tend to involve multi-disciplinary teams, with specific expertise in the management and manipulation of relevant data. While individual skills remain important – communication, collaboration, persuasion and integration can be seen as the essence of functional team performance. It is acknowledged that one individual supremo is unlikely to encompass all the knowledge and insight necessary to solve complex issues. This may have always been the case to some extent, but the acceleration in project complexity makes it crucial to project success in today's world. The hero architect or hero builder can bring an unhealthy degree of ego to a project team. Of course, this is not always the case, but there is a widely held perception that ego plays a significant role in the minds of many designers. Yet, in the past, it has also led to a clear vision for the design. Although not precluding the possibility of an individual team leader, it is essentially true that teams are becoming the primary unit of work organisation in design and project delivery. The role of teams, what makes a successful team, and how team performance can be enhanced is a widely studied field in project delivery (Senaratne and Gunawardane 2015). Team members tend to naturally adopt roles to suit their functional roles. This process can be enhanced by those responsible for team formation. A balanced team is the aim of clients and project managers alike, but this can be a difficult task when the proposed design project contains many unknown quantities. Leadership style is frequently critical to a well-functioning team. Good leaders bring out the best in other team members and do not rely on their position alone for their authority. Instead, they gain respect for their capacity to bridge disparate information sources and make justifiable choices. Additionally, there is evidence that leadership diversity assists in tackling issues that involve human resource management and integrated project delivery (Baker et al. 2021). Larger companies throughout the Architecture Engineering and Construction sector are currently actively seeking to diversify both team membership and leadership in their teams. This aims at addressing inherent cultural biases as well as finding newer creative insights that may come from different life stories and diverse backgrounds of the team members. In particular, the industry in places like Australia is actively encouraging the indigenous population, which has been long under-represented in the design and construction management professions (Go-Sam 2022). Under-representation may stem from historical issues of many kinds, but it is increasingly recognised that more culturally representative teams make better decisions.

Team balance, team diversity and evolving expert knowledge are significant inputs to produce a more agile response to design problems. This balance includes professional, social, gender, ethnic and linguistic dimensions. Personality type might also be considered as a dimension of team formation. The comparatively new professional roles of the design manager or BIM manager will need to find an appropriate place in the decision-making structure rather than duplicating or overseeing work already done by the architect. This adds to concerns about managing design quality. Yet, the benefits of socially diversifying the team will be lost if the team finds itself in an old-school, rigid hierarchical structure. Such a structure would be seen as culturally inappropriate in many indigenous societies and in many other cultures outside the developed world (Potter 2012). A significant aspect of cultural assumptions is that, by nature, they are not easily recognised by the holders of the assumptions. This is often a major source of resistance to team diversification. Human beings often favour hearing their own beliefs and ideas repeated back to them. Opposition to diversity can be described in terms of meritocracy when it is really about the inability to value viewpoints other than one's own. Consequently, dispute resolution in teams is critical to successful outcomes. Creating a culture of respectful listening as a pre-emptive strategy for dispute resolution is likely to have positive outcomes for all involved. Also, inclusive decision-making has been demonstrated to contribute to more sustainable outcomes in urban development (London and Cadman 2009). In addition, team balance needs to consider the sometimes disparate skills needed to appreciate the dimensions of economic, social and environmental sustainability. It can be argued that champions for each of these three dimensions are needed in the design team for any complex project. Reasoned discussion and evidence-based decision-making speak for themselves as essentials in this inclusive design process. Cultural awareness can mediate against the tendency of some individuals to be reluctant to disagree with the nominal team leader. Teams need to be clearly authorised to have open discussions and to raise issues of importance that may otherwise be overlooked. Without this explicit authorisation to question, real inclusion is less likely to occur.

Multi-faceted communication skills are the essential lubricant for well-functioning teams. In addition to empowering all team members to speak up, design team members need to be equipped with the vocabulary, graphical and empathetic techniques that enable persuasive arguments for the most effective project solutions. Some team members may need assistance via a facilitator or translator for these matters. Consensus building within a team faced with multi-faceted tasks is necessary, but it should not involve groupthink, which typically results in unchallenged, poor-quality decision-making. Rather, consensus should grow in an atmosphere where all team members are empowered to speak and to listen. This has been found to lead to better decision-making and more successful outcomes (Singhaputtangkul and Xianbo 2016). Group cohesiveness based on good communication is likely to be much more effective than cohesiveness based on adherence to a pre-established hierarchy. Negotiation between the various professions involved in building project delivery has not always been helpful to the consensus-building process. Increasingly, it is the relevant experience of the participants that is seen to give weight to their role in a project delivery team as opposed to their formal qualifications. This is because of the nature of building design being essentially a prototype for each building project. Architects have usually taken this position and increasingly use their 3D computer visualisation skills to improve intra-project communication. Moore (2012) describes this process in relation to the successful delivery of a Melbourne academic building. In commercial building projects. However, disputes and problem-solving issues can often centre around cost and time rather than clear visualisation of the completed building. Shared values within the team make consensus more likely, but it should be remembered that a diverse project team may not have strong intersections between all the members on issues of innate value or correct action. Ethical behaviour within teams needs to be specifically mandated in the formation of teams from the commencement of the endeavour.

While personal skills and leadership styles have an enormous impact on team coherence and problem-solving, it should also be remembered that technology is changing the very nature of our interaction with others and our approach to the critical issues of our age. Digital natives who have always lived within this context will almost inevitably perceive things differently from others. The ongoing digital transformation generates possibilities but also some significant challenges for different contexts. New leadership schemas that seek to integrate digital technology in new ways are increasing the readiness for change and the capacity to manage change by those who have access to it (Criado-Perez et al. 2022). Essentially, the masters of the new digital tools have the inside running in the leadership stakes for new and complex projects. This may be difficult for those individuals who have a long history of non-digital practice and who have developed their design expertise in the analogue world and continue to live within it. Well-known Pritzker Prize architect Glenn

Murcutt is widely reported to eschew digital design tools. It is unlikely that this choice will be open to future generations of designerarchitects for large and complex projects. The potential power of digitalisation is so great that it may subsume the capacity of an individual, albeit an enormously talented one, to deliver design solutions that can be supported by data and informed by parametric design. As a consequence, the design projects of the future are likely to be heavily based on broad data collection and quantitative analysis rather than solely on individual intuitive design inspiration. This can be seen as a positive development because of the potential to absorb more real-world observation and contribute to a better understanding of genuine sustainability in the built environment. Of course, change often generates corresponding losses alongside the gains. There are losers as well as winners in the change process. Emphasis on data may move building design away from traditional techniques towards manufacture, with the international masters of the new digital tools supplanting local analogue designers.

Ostwald (2010) has studied ethical decision-making in the generative design process. He finds that the growing use of sophisticated software has enabled designers to experiment with biomorphic and topographic forms in a way not previously possible. It was found that clarity of authorship, appropriate care and attention and designer motivation were all helpful when determining ethical merit of design. Transferring these values to the design team environment can provoke differing viewpoints and complex negotiations. Disagreement is not a problem in itself, provided there is open discussion and fair resolution of disputes. Architectural spatial design is recognised as a complex process that usually has many potential solutions, and often it is difficult to determine the optimal solution, without substantive experience (Campbell et al. 2014). Generative design is achieved through parametric and algorithmic processes. While this concept can be challenging for an experienced architect, it raises the possibility of supplementing the designer's knowledge with real-time feedback on spatial decisions made during the concept design phase. Campbell et al. (2014) endorse parametric modelling as a parallel stream to traditional design generation from structured precedent analysis. The potential solutions that can be tried represent a quantum leap in spatial design analysis. The use of parametric design is also endorsed by Lee et al. (2014) as a way of better serving the needs of the construction industry in terms of effective adaptation of existing knowledge and visualisation capacity. Project delivery teams can take advantage of the potential of parametric design techniques to genuinely democratise choices and solutions across a range of building projects. Design cognition, which has often been seen as a barrier to more inclusive decision-making, can be enhanced by parametric design to foster the emergence of a shared team model and a focus on goal-directed behaviour in teams (Dong et al. 2013). Given the appropriate tools, generating designs could be open and available for all to experience and not reserved exclusively for professionals. However, this raises concerns that the skills of architects may be underrated and that anybody considers that they can design buildings because digital tools offer the illusion that they can.

The notion of wicked problems was first broached by Rittel and Weber in the 1970s (Rittel and Webber 1973). The identification of this kind of intricacy assists in the understanding the need to be multi-faceted in addressing these issues. The concept has been revisited several times and has been seen as particularly relevant to the design process with its innate complexity and multi-disciplinary nature (Farrell and Hooker 2013). The authors dispute the claim that generic design is an endeavour separated from science and that the two cannot be reconciled to any notable extent. On the contrary, they describe a need for boundary-spanning to bring insights from diverse perspectives. Synergies are described between the design process and scientific exploration and the generation of theory. These synergies can create a shift towards strategic problem-solving, which opens new possibilities for cross-disciplinary research and practice. Architectural design has always been perceived as the interaction of arts and sciences. Additionally, Crowley and Head (2017), in their careful revisiting of the ideas of Rittel and Webber (1973), describe how these issues, their scope and pathways towards resolution remain important in the sociopolitical context of the 21st century. As a result, innovative structures and evaluation methods could be needed to meet broader challenges. Economic, environmental, social and governance parameters should be explicitly addressed in contemporary design processes. It is certainly arguable that the data-based procedures of digital tools may not be best placed to represent this kind of approach.

Change Agents

It has already been shown that new roles within project teams are taking increasing prominence, especially in larger and more complex projects. The nomenclature varies throughout the world and across languages and cultural clusters. However, these new roles are more than just job titles, as they signal changes in the distribution of responsibility, as well as the authority to be the ultimate decision-makers and arbiters on projects. The prevalence of largely competency-based titles such as BIM manager, BIM project manager, BIM coordinator and BIM technician is an indicator of accelerating change (Uhm et al. 2017). These titles tend to be very loosely defined. They are also differently applied across countries, projects and organisations. Nevertheless, they represent markers of evolution in the actual process of project delivery. Not all changed roles are driven by technology. There is also change stemming from the educational pathways that feed individuals into these roles. Emmitt (2016) chronicled the rise of the construction design manager in the early 21st century from a job description that was virtually unknown during the previous century to a booming career path, especially amongst design and build organisations. The barrier between designers and makers is becoming increasingly fuzzy in certain areas. Agile and perceptive individuals can clearly span both roles. Design managers may have started as document coordinators, but increasingly they have become involved with issues of quality, environmental compliance and buildability. Their role has evolved into a more proactive one and has resulted in a push for earlier involvement in the design and preconstruction phases. They have also been reported as playing the role of defenders of the design intent against inappropriate cost-cutting. In addition, some design managers have been lobbying for better quality in the suitability and completeness of the documentation supplied by design teams.

It is paradoxical that the increasing documentation power provided by BIM capabilities may have led to a diminution in the expression of design quality in the drawings. Perhaps, it could be due to a mismatch between the technological competency of computer-aided design and practical building knowledge. This perceived deficiency could increase over time as educators and trainers place more emphasis on delivering solutions that can be fabricated rather than viewing the outcome as architecture composed of form, space and structure. Perhaps unfairly, architects in the digital arena have sometimes been accused of focusing on the virtual model rather than the proposed building. In fact, with a balanced education, design thinking and visualisation can be virtual change agents in the delivery of quality projects provided there is not a disconnect between what happens in the office and the output on the building site (Rowlinson and Yates 2003). All team members need to acknowledge the value of information from previous projects as contributions to design solutions. Before the advent of information technology, it could be difficult to review this information from paper-based archives. With the benefit of BIM tools, a genuine change agent can facilitate incorporating previously imponderable amounts of detail into the design decision-making process.

Change agents may be technological advances in themselves as much as human beings with new ideas. The ubiquity of digital devices in today's world affords the potential for people at all levels of education and training to embrace the use of IT to improve efficiency and accuracy in their work choices. Mobile phones have infiltrated almost all regions of the populated world, and younger generations find the connectivity that is generated by this one device indispensable in their lives. While once it would have seemed very unlikely that tradesmen and other construction workers would be persuaded to rely on digital communication in their working lives, it is now almost completely incorporated into most workspaces, including construction sites. This change is not confined to the developed world. The mobile phone is the digital device that paved the way for tools such as iPads and similar portable devices to become essential tools, gradually replacing clipboards and paper documents. Mäki and Kerosuo (2015) report on the gradual penetration of BIM and related software into the site manager's role. At first, tentative and limited by the nascent competencies of the site managers in the digital field, this is now accelerating. The competitive advantage gained by early adopters is driving a gradual digitalisation of construction delivery across a broad spectrum of companies and project roles. When this competence is fully achieved across the population of those involved, it is possible for a new ecosystem of digitally aware individuals to coalesce into new integrated teams where roles are not determined by professional labels but by the capacity to communicate with one another, as a means of moving towards new solutions to environmental, social, economic, political and cultural dilemmas. In other words, the tools are powerful in themselves, but the users of the tools can integrate their practice across different disciplines and are potentially even more powerful as change agents. Nevertheless, it needs to be remembered that tools cannot replace skills and knowledge. Whilst acknowledging the contributions of other team members, architectural education is a necessary prerequisite to designing buildings.

Specialised uses of BIM technology are developing to address long-term issues where previously qualitative and quantitative data gathering was not considered practical. The data may have been on a scale that made non-digital analysis prohibitively time-consuming and not amenable to thoughtful analysis. Just a few examples will suffice for illustration. Firstly, Bastem and Cekmis (2022) report that BIM is being used for collaborative management of heritage buildings in ancient but still developing cities in Turkey. They describe how heritage building information modelling (HBIM) can intelligently document, interpret and manage the culturally significant built heritage of a city. The authors foresee

HBIM replacing traditional survey and restoration processes in the near future. Secondly, BIM is reportedly being used to facilitate deconstruction at the end of a building's lifespan by easily identifying reusable elements and tagging them in the digital model, as with the legacy of any fully documented BIM project (van den Berg et al. 2021). Once again, this is addressing an issue that existed long before the digital revolution but can be tackled systematically with the aid of real data in virtual models. Thirdly, although the adoption of BIM has long been criticised as not generating actual cost efficiencies, some very large data studies show the cost–benefit analysis is significant (Chahrour et al. 2021). It is likely that few large and complex projects in the future will not incorporate the use of virtual modelling because the potential advantages for these types of projects are demonstrable. Indeed, much of the education sector in the developed world is actively pursuing upskilling at all levels as an acknowledgement that digital capacities are critical for the jobs of the future (Ariyachandra et al. 2022).

The concept of liminality can be described as referring to adaptation to change, across boundaries, between one role and the next. Gustavsson (2018) uses this concept derived from anthropology to study new professional roles in the management of building projects that challenge existing established structures and communities of practice. Negotiation in the boundary spaces between former defined professional specialities can be described as liminal experiences. These experiences generate tensions and uncertainties among team members. New industry roles such as partnering manager and building logistics specialist can be viewed as part of a change mechanism that requires communication and mutual understanding if they are to be effective (Gustavsson 2018). Using the lens of liminality to observe and describe the process of change yields the insight that the dynamics of change frequently result in tensions and disputes because individuals within the process constantly feel that they are working within a fluid space between established practice and an emergent reality. These tensions result from multiple interactions along the boundary spaces involving different paradigms. This theoretical explanation of the lived experience of many in a changing society is helpful to an understanding of how and why change proceeds or fails to proceed. Individuals who see themselves as change agents would benefit from acknowledging the difficulties experienced by those who are caught up in a change environment and who may lack agency to control the outcomes for themselves and others. Whether architects or builders or even data managers will prove to be the ultimate boundary spanners is not yet known, but the challenge to all current actors in the built environment is to look into the future with minds open to externally generated solutions.

Facilitated Collaboration

In the context of widespread change and increasing complexity, it is essential that those who establish teams to deliver building projects focus on facilitating collaboration within diversity. This collaboration needs to be founded on deep understanding rather than tokenism. In design teams, collaboration requires individual collaboration skills as well as individual discipline skills (Kleinsmann et al. 2012). Knowledge-sharing and knowledge-integration are needed in addition to the individual design skills, and there is often a non-linear relationship between the different skills. There is an essential generosity involved in teamwork, which must acknowledge diverse contributions. This is well-recognised in fields such as education and the arts but can be particularly difficult to reproduce in commercial project delivery teams throughout the built environment. Personal attributes are needed in the team that include deep listening, genuine inclusion of other perspectives, team ethos building and clarity of expectations (Rahimian et al. 2022). Deep connections among project stakeholders, alongside high-quality communication on the issues of disagreement, can improve outcomes for all concerned. Disputes in the process of project delivery can often centre around team members' rival concepts of buildability (Nibbelink et al. 2017). This has been a long-standing source of disagreement. Digital tools available today can assist with genuine understanding of these issues by means of fuller visualisation and the incorporation of time frame parameters in the design and detailing process. The separation between design and making had its roots in 19th-century ideas about social class but can potentially be reversed by the capacity of participants with diverse skills to be involved in modelling a future project and sharing their tacit expertise gained from previous project delivery. The ultimate outcomes will have a positive effect from the contributions of all participants.

The theoretical construct of absorptive capacity attempts to measure the capacity of an organisation to incorporate external knowledge into its processes and procedures. A recent study of building contractors' absorptive capacity, with respect to building information modelling, found that some of the potential benefits of BIM could be to give contractors greater ability to understand design intent and facilitate project integration (Ahankoob et al. 2021). In fact, it was found that the use of BIM could aid transparency in project relations and support competitive advantage for BIM-savvy contractors. Many of them are adopting the notion of best-for-project as a governing principle, which can foster cooperation rather than competition within a project team. Clients and developers as the originators of many projects can drive the move towards team collaboration by making it part of the initial construct of their project teams. Design risk is a critical part of design-build project procurement. Time-critical inputs to projects are often the focus of experienced building contractors and their specific areas of expertise. Architects tend to have less knowledge in this area and less concern about accurate time and workflow forecasts. The risk of inaccuracy is a significant aspect of design and build procurement and is also the source of contractor suspicion about integrated project teams. Mitigating design risk is a major driver for change in projects (Junying et al. 2017). Suspicion and confrontation between rival team members are likely to continue unless the digital expansion enables full cooperative integration between the design and project delivery teams.

Design for Manufacture and Assembly is a comparatively recent strategy for integrating the processes of designing and building. It represents both a philosophy and a methodology for integrating supply chains that combine offsite manufacture with onsite assembly (Gao et al. 2020). This process is gaining momentum in many parts of the world and has received impetus where recent climate change-driven natural disasters (fire, flood and storm) have underlined the need to be able to quickly replace the dwellings, workplaces, schools and health facilities that are lost due to such natural disasters. In addition, the experience of the pandemic has led many people to rethink some of the globalisation outcomes of the past few decades. The question of sovereign and independent supply chains is now being raised by many commentators. The adoption of greater levels of DfMA illustrates all the team communication issues and potential misunderstandings that have been discussed here in the wider context of building procurement. The potential for solutions that serve both social and economic purposes is great, but the role of integrators from different perspectives is still problematic. It is likely that expertise in logistics will become increasingly important. New jobs in the delivery of projects which are largely built offsite will challenge the existing industry as well as the training and education sector. The challenge of delivering a more sustainable built world in towns, cities and regions is one that can be contributed to, by greater use of offsite manufacturing technologies. These new offsite jobs can also potentially be better, cleaner and safer jobs for the people involved. However, the re-emergence of this kind in industrial production will limit architectural language and response to locality.

Ways Forward

Insights from team formation theory can aid in the establishment of a more flexible and multi-disciplinary project delivery environment for the reworking of the existing built environment in a more sustainable manner. Diversity in team formation across a wide spectrum of criteria is vital to the composition of teams that can fully represent the end users of projects. This diversity is not limited to variety in the professional backgrounds of the team, although that is clearly needed. Project teams also need to be culturally, gender-, age and linguistically diverse and to incorporate people of different abilities and different life experiences. The plethora of digital tools that are increasingly available can be part of the solution. The ability to manage and manipulate data towards validating design solutions is a critical skill. Inclusive and distributed decision-making that leads to integrated project design and delivery has been shown to contribute to outcomes for the end users of the built environment. Facilitated knowledge-sharing and teams that focus on outcomes and measurable benefits are beneficial to further progress in performance attributes, although there remain issues in terms of amenity attributes (see Figure 12.1).

Managerially based design teams, or design and build teams, together with digital tools, may well homogenise architectural language. The predominant image of the built environment could be generated by artificial materials and systems. They may diminish a sense of place and local characteristics. In some parts of the world, systems may be imposed on design communities that do not possess the sophisticated digitalisation needed to enact them. In this way, they may lose moral authority over their own constituencies. Nevertheless, mutual respect and the capturing of tacit knowledge for broader benefit are important strategies and have generally occurred in traditional practice. In certain locations, future directions in the built environment could depend on the development of a new paradigm that supports rather than obscures common interests. There is a need to promote inclusion proactively at all levels in a variety of contexts.

Some of these perspectives may seem almost impossible to reconcile. It is also important to remember that towns and cities have invariably arisen out of a level of chaos and disorganisation where positive actors have sought to make them more orderly. Perhaps more than at any time in human history, there is an opportunity to harness the insights and skills of large and diverse populations to address the critical issues of the age. There is not the same kind of geographical obscurity that was experienced by previous generations, as air travel has brought places closer together and global communication networks can transfer

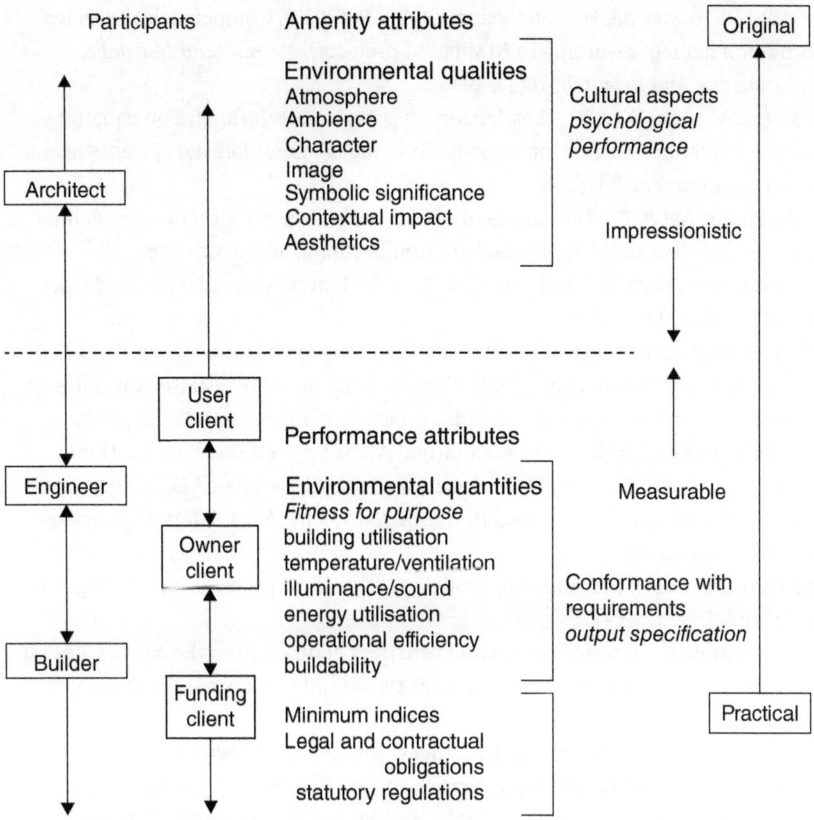

Figure 12.1 Design quality hierarchy (Giddings and Melhado (2019) with permission of CIB General Secretariat).

information instantly. There is potential access to a vast array of stored experience from past generations and tools to harness and analyse information for the benefit of present and future generations. The process of change is often problematic for those who experience the tensions of transition. Insights from liminal theory can inform strategies to deal with these issues. For those who are privileged to play a role in the delivery of the built environment, there is a corresponding obligation to acknowledge the insights and experiences of others who share the planet. It is important to cast nets widely if the benefits of these exciting times are to be equitably shaped and shared by different populations in this extraordinarily diverse world.

References

Ahankoob, A., Manley, K., Hon, C. and Drogemuller, R. (2021). The influence of building information modelling on the absorptive capacity of project-based organisations. *Architectural Engineering and Design Management*. https://doi.org/10.1080/17452007.2021.1881879.

Ariyachandra, M.R.M.F., Jayasena, H.S. and Perera, B.A.K.S. (2022). Competencies expected from an information manager working in BIM based projects. *International Journal of Construction Education and Research*, 18(1), 49–66.

Baker, M., Ali, M. and French, E. (2021). Leadership diversity and its influence on equality initiatives and performance: insights for construction management. *Journal of Construction Engineering and Management*, 147(10), 1–12.

Bastem, S.S. and Cekmis, A. (2022). Development of historic building information modelling: a systematic literature review. *Building Research and Information*, 50(5), 527–558.

Campbell, M.I., Shea, K., Coorey, B.P. and Jupp, J.R. (2014). Generative spatial performance design system. *AI EDAM*, 28(3), 277–283.

Chahrour, R., Hafeez, M.A., Ahmad, A.M., Sulieman, H.I., Dawood, H., Rodriguez-Trejo, S., Kassem, M., Naji, K.K. and Dawood, N. (2021). Cost-benefit analysis of BIM-enabled design clash detection and resolution. *Construction Management and Economics*, 39(1), 55–72.

Criado-Perez, C., Shinkle, G.A., Höllerer, M.A., Sharma, A., Collins, C., Gardner, N., Hank Haeusler, M. and Pan, S. (2022). Digital transformation in the Australian AEC industry: prevailing issues and prospective leadership thinking. *Journal of Construction Engineering and Management*, 148(1), 1–12.

Crowley, K. and Head, B. (2017). The enduring challenge of 'wicked problems': revisiting Rittel and Webber. *Policy Sciences*, 50(4), 539–547.

Doloi, H. (2008). Analysing the novated design and construct contract from the client's, design team's and contractor's perspectives. *Construction Management and Economics*, 26(11), 1181–1196.

Dong, A., Kleinsmann, M.S. and Deken, F. (2013). Investigating design cognition in the construction and enactment of team mental models. *Design Studies*, 34(1), 1–33.

Elghaish, F., Pour Rahimian, F., Hosseini, M.R., Edwards, D. and Shelbourn, M. (2022). Financial management of construction projects: hyperledger fabric and chaincode solutions. *Automation in Construction*, 137. https://doi.org/10.1016/j.autcon.2022.104185.

Emmitt, S. (2016). The construction design manager – a rapidly evolving innovation. *Architectural Engineering and Design Management*, 12(2), 138–148.

Ershadi, M., Jefferies, M., Davis, P. and Mojtahedi, M. (2021). Comparative analysis of PMO functions between the public and private sectors: survey of high-performing construction organizations. *Journal of Construction Engineering and Management*, 147(11), 1–16.

Farrell, R. and Hooker, C. (2013). Design, science and wicked problems. *Design Studies*, 34(6), 681–705.

Ferme, L., Zuo, J. and Rameezdeen, R. (2018). Improving collaboration among stakeholders in green building projects: role of early contractor involvement. *Journal of Legal Affairs and Dispute Resolution in Engineering and Construction*, 10(4). https://doi.org/10.1061/(ASCE)LA.1943-4170.0000278.

Gao, S., Jin, R. and Lu, W. (2020). Design for manufacture and assembly in construction: a review. *Building Research and Information*, 48(5), 538–550.

Go-Sam, C. (2022). Indigenizing practice: Maximizing economic benefits. *Architecture Australia*, 111(3), 13–16.

Giddings, B. and Melhado, S. (2019). *Architectural Design and Management Research Roadmap*. CIB Publication 416. Ottawa: CIB General Secretariat.

Gustavsson, T.K. (2018). Liminal roles in construction project practice: exploring change through the roles of partnering manager, building logistic specialist and BIM coordinator. *Construction Management and Economics*, 36(11), 599–610.

Junying, L., Qunxia, X., Bo, X. and Bridge, A.J. (2017). Impact of design risk on the performance of design-build projects. *Journal of Construction Engineering and Management*, 143(6), 1–10.

Kleinsmann, M., Deken, F., Dong, A. and Lauche, K. (2012). Development of design collaboration skills. *Journal of Engineering Design*, 23(7), 485–506.

Langston, C. and Zhang, W. (2021). DfMA: towards an integrated strategy for a more productive and sustainable construction industry in Australia. *Sustainability*, 13(16). https://doi.org/10.3390/su13169219.

Lee, J., Gu, N. and Williams, A.P. (2014). Parametric design strategies for the generation of creative designs. *International Journal of Architectural Computing*, 12(3), 263–282.

London, K.A. and Cadman, K. (2009). Impact of a fragmented regulatory environment on sustainable urban development design management. *Architectural Engineering and Design Management*, 5(1/2), 5–23.

Mäki, T. and Kerosuo, H. (2015). Site managers' daily work and the uses of building information modelling in construction site management. *Construction Management and Economics*, 33(3), 163–175.

Manley, K., McFallan, S. and Kajewski, S. (2009). Relationship between construction firm strategies and innovation outcomes. *Journal of Construction Engineering and Management*, 135(8), 764–771.

Moore, T. (2012). Reflection. *Architecture Australia*, 101(5), 10.

Nibbelink, J.-G.Sutrisna, M. and Zaman, A.U. (2017). Unlocking the potential of early contractor involvement in reducing design risks in commercial building refurbishment projects – a Western Australian perspective. *Architectural Engineering and Design Management*, 13(6), 439–456.

Nolan, G. (2018). Managing risk while translating research outcomes into design and construction innovation. *Architectural Science Review*, 61(4), 255–65.

Ostwald, M. (2010). Ethics and the auto-generative design process. *Building Research and Information*, 38(4), 390–400.

Potter, E. (2012). Introduction: making Indigenous place in the Australian city. *Postcolonial Studies*, 15(2), 131–142.

Rahimian, A., Hosseini, M.R., Martek, I. et al. (2022). Predicting communication quality in construction projects: a fully-connected deep neural network approach. *Automation in Construction*, 139. https://doi.org/10.1016/j.autcon.2022.104268.

Rittel, H. and Webber, M. (1973). Dilemmas in a general theory of planning. *Policy Sciences*, 4(2), 155–169.

Rowlinson, S. and Yates, D. (2003). nDCAD: a virtual change agent for professions and procurement systems?*Construction Management and Economics*, 21(8), 849–57.

Senaratne, S. and Gunawardane, S. (2015). Application of team role theory to construction design teams. *Architectural Engineering and Design Management*, 11(1), 1–20.

Shen, Y.J. and Walker, D.H.T. (2001). Integrating OHS, EMS and QM with constructability principles when construction planning – a design and construct project case study. *The TQM Magazine*, 13(4), 247–259.

Singhaputtangkul, N. and Xianbo, Z. (2016). Applying a fuzzy consensus scheme to enhance the group decision making of a building envelope design team. *Journal of Construction Engineering and Management*, 142(8), 1–11.

Uhm, M., Lee, G. and Jeon, B. (2017). An analysis of BIM jobs and competencies based on the use of terms in the industry. *Automation in Construction*, 81, 67–98.

van den Berg, M., Voordijk, H. and Adriaanse, A. (2021). BIM uses for deconstruction: an activity-theoretical perspective on reorganising end-of-life practices. *Construction Management and Economics*, 39(4), 323–339.

Zou, P.X.W., Redman, S. and Windon, S. (2008). Case studies on risk and opportunity at design stage of building projects in Australia: focus on safety. *Architectural Engineering and Design Management*, 4(3/4), 221–238.

13

The Language of Digital Technologies

Leonardo Manzione[1] *and Silvio Melhado*[2]

[1] *Coordenar, Sao Paulo, Brazil*
[2] *University of Quebec, Montreal, Canada*

Introduction

The French linguist Benveniste (1971) said that language is a system of socialised signs, and its elements only acquire meaning when inserted into contexts since each sign alone does not have a relevant meaning. Therefore, the concepts of language and culture do not exist in isolation, as each culture, each era and each technology builds and articulates its own language to meet different interests and needs. This is a current situation in times of big data and the flood of information provided by the content of new digital technologies. These technologies have created a diverging vocabulary, thus hindering common understanding. Two major elements are used for communication to fully materialise. They are language that represents an entire system of conventional signs, whether verbal or non-verbal, and language that represents a system of conventional grammatical signs used by the members of a particular community. Following this presupposition, the linguistic sign is conceived as a representative element consisting of two basic aspects: the signifier and the signified. Any object, sound or word capable of representing something else can be said to constitute a sign. People depend on signs to live and to interact with the environment in which they are found. For humans, the notion of signs and their relationships is not so important from the theoretical point of view, yet they are understood in a practical and precise manner. The interface between objects and signs, and thus the signs themselves, can be defined functionally since signs are employed to signify the objects. Meier-Oeser (2011) evokes a definition of a sign that has been attributed to Saint Augustine, which is that a sign is received by the senses but conveys something else to the intellect. The sign can be seen as the atom of language, as it is the fundamental unit for understanding a code and may be divided into two levels of understanding:

- Signifier: is the tangible, perceptible, material element of the sign.
- Signified: is the concept, the abstract entity of the sign.

Architectural Design and Management in the Digital Age: International Perspectives, First Edition.
Edited by Bob Giddings, Silvio Melhado, and Jenni Barrett.

The linguistic sign is the fusion of a sound pattern and a concept of a signifier and a signified. This fusion, in turn, is neither determined nor fixed, but rather arbitrary. Barthes (1964) adds that the signifier constitutes the narrative order, and the signified the content order. He defines the signified as the psychic representation rather than the object itself. The signifier may be analysed with the same observations placed for the signified, the only difference being that the signifier of a mediating element behaves as a generator i.e. the materialiser of the object figure, the signified. Lastly, Barthes (1964) adds that the signifier constitutes the narrative order, and the signified the content order. Additionally, the linguistics of De Saussure (2011) works on two fundamental principles in the analysis of the sign: (i) a principle of arbitrariness, since the line that binds signifier and signified is not determined from a place outside the language and there is no need to unite a given signifier to a given signified; and (ii) a principle of linearity of the signifier, since its nature is auditory and its expression, whether in speech or in the mind, is given by the consecutive emergence of one sign after the other in the timeline. Discussing the nature of the linguistic sign is essential for de Saussure (2011), as he opposes the traditional view that language is a classification system. Were it such a system, language would be a list of terms corresponding to the items in another list, which places ideas at a level previous to words since they would merely have the function of nominating ready, fixed ideas. As a result, there is the view that an idea comes before a word and, therefore, the word exists as a result of the idea but is not a part of its genesis. For de Saussure (2011), a linguistic sign is not a link between an object and a name, as he believes that actual objects and names do not matter in this context, but rather between a concept and a sound pattern. He states that the sound pattern is not actually a sound, for a sound is something physical. A sound pattern is the hearer's psychological impression of a sound, as given to him by the evidence of his senses. This sound pattern may be called a material element only in that it is the representation of our sensory impressions. The sound pattern may thus be distinguished from the other element associated with it in a linguistic sign. This other element is of a more abstract kind, ie the concept. (De Saussure 2011). The nature of sound patterns is thus psychic, not physical. It is not the sound that matters, but an acoustic, sensory impression that emerges in an individual's mind when, for example, without uttering it, one thinks of a word. In sum, the linguistic sign is a two-faced psychic entity, represented by Figure 13.1.

This set of concepts and sound patterns could also be referred to as a sign. De Saussure (2011) thus names concept and sound pattern as signified and signifier, respectively, with a pedagogical function. Neither signified nor signifier are united by external determination. As noted above, what binds them is arbitrary. Hence, the sign itself results from this union and can also be classified as arbitrary. The word arbitrary requires further explanation. It should not give the idea that the signified depends on a free choice of what is uttered. The

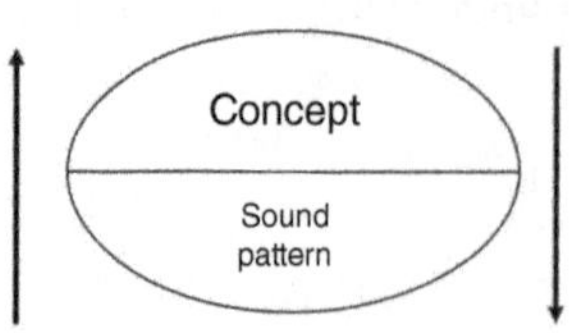

Figure 13.1 Linguistic sign representation (Adapted from De Saussure 2011).

point is that the signifier is unmotivated and that it is not connected with the signified, with which it has no natural link (De Saussure 2011). A symbol does not express the same arbitrariness as the sign. It contains within itself the root of an external link with language. The symbol of justice, for example, the balance scales, cannot therefore be replaced by any other object since the union between signifier and signified has a direct bond in history, tradition and symbolism. The signifier is of auditory nature and is thus developed in a timeline. It presents an extension and can only be measured in a sense and in a single direction, as it is a line. Therefore, the signifier is both arbitrary and linear in character.

BIM, a Sign Abounding with Signifieds and Signifiers

The notion of an information model is not bonded by any intrinsic relationship to the sequence of sounds i-n-f-o-r-m-a-t-i-o-n m-o-d-e-l that serves as its signifier. It could equally well be represented by another sequence, such as Building Information Modelling (BIM), which will be discussed in depth, as proof that there are differences between languages and the very existence of different languages. Concurrently, the same signifiers are noticed to be often used to express the same signified in similar situations. This fixity is of social nature and exists because language is practised collectively. The existence of stable terms does not nullify the first principle of the order of signs, i.e. the principle of the arbitrariness of the sign. The acronym BIM is a powerful sign in the language of digital technologies. From the linguistic viewpoint, according to the theory presented in the previous headings, BIM as a sign presents an abundance of signifieds and signifiers. This abundance may result in disputes and contractual issues related to the scope of the work and to the deliverables themselves, which may be justified by different and diverging understandings of the term BIM. Gaspar's (2019) work focus is to temporally and referentially qualify the understanding of the signifieds attributed to the Building Information Modelling term and related ones. The oldest definition of the term building information model found by Gaspar (2019) is that of Nederveen and Tolman (1992). They proposed that a building information model be composed of different aspect models so that it is possible to visualise and to decompose its parts whenever desired. The oldest definition of BIM found by Gaspar (2019) is that by Fu et al. (2004). Their aim was to show the potential of implementing methods and techniques of BIM by using international foundation class models as a scheme for exchanging data in developing cost-estimating tools for the entire lifecycle of a building construction. As an important result of his research, Gaspar (2019) verified the need for there being three groups for categorising understanding related to the different BIM signifieds and signifiers. These are models, processes and systems. In his research, Gaspar (2019) proposed that the BIM sign simultaneously has four signifieds and two signifiers, as shown in Table 13.1.

With the first signifier, the model and its signified model applications, Gaspar (2019) verified that all the definitions proposed between 1989 and 2014 are similar, thus indicating a consensus among researchers as regards the signified of the term. This relates to the production and application of models to conduct different analyses for all the stages of a development lifecycle. In these definitions, there is a predominance of a signified attributed to the sign with a bias connected to the application of development digital models, according

Table 13.1 BIM signifiers and signifieds (Gaspar 2019).

Sign	
BIM	
Signifiers	**Signifieds**
Model	Model applications
Modelling	Management processes Computer systems High-performance projects

to the objectives of each agent. With the second signifier, modelling and its first signified, management processes, BIM showed an evolution of the signified predominantly attributed to the signifier. It migrated from the technological axis towards themes related to processes and to the management of all the activities structuring the civil construction sector, originating the definitions that expand the application of the term beyond the strictly computational and technological domains, taking them to the boundaries of the business and innovation management processes. The second signifier and the second signified, Computer Systems, in 2002, Autodesk developed the use of computer systems that aid the development of projects, providing users with tools that stimulate them to design with the object-oriented modelling method (Gaspar 2019). With the second signifier and the third signified, high-performance projects, as a corollary of his research, Gaspar (2019) identified a latent signified attributed to BIM from the viewpoint of the scientific community and of the market. He proposed that BIM would be an adequate term to define architectural and engineering solutions aimed at producing high-quality data for developing projects with a high-performance level. He concludes by proposing a new signified as "Building Information Modelling defines the data production process for high performance designs" (Gaspar 2019:8).

From PAS 1192 to ISO 19650: A History of Findings, Promises and Language

In 2010, the International Council for Research and Innovation in Building and Construction (CIB) identified Integrated Design and Delivery Solutions (IDDS) as a priority theme, through the framework for an integrated and coordinated merge of people, process and technology issues in order to conduct a radical and continuous transformation in design and construction processes. Prins and Owen (2010: 227) note that "IDDS will use collaborative work processes and enhanced skills together with integrated data, information, and knowledge management to minimise structural and process inefficiencies and to enhance the value delivered during design, build, operation, and across projects. IDDS build on building information modelling (BIM), incorporating advances in the training and employment of people, together with supporting new technologies." However, the process had started before then. In the United Kingdom, a sequence of initiatives initiated during

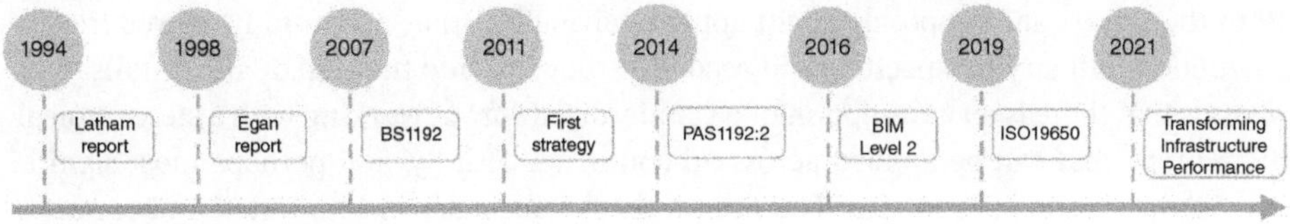

Figure 13.2 UK BIM timeline.

the 1990s were initiated to achieve an improvement in productivity in the construction industry. Becoming very influential in many other countries, it is important to mention that the Latham Report and the Egan Report stressed the need for dramatic changes to achieve more profitable outcomes (Fernie et al. 2006).

A good way of presenting this history is through a timeline represented in Figure 13.2.

The following is a brief description of the events occurring in the timeline, seeking to highlight the milestones of BIM in the United Kingdom. The Latham Report *Constructing the Team* was commissioned by the British government to investigate the issues perceived in the construction industry, which Sir Michael Latham, the author of the report, described as adversarial, ineffective, fragmented and incapable of delivering for its clients (Latham 1994). He proposed that the client should be at the centre of the building construction process and that the industry should move away from its adversary structure, adopting a more integrated approach, with greater partnership and teamwork. He suggested that, if the full range of measures described in the report were adopted, a 30% saving could be derived in five years. The Latham report led to the establishment of the Building Industry Council (BIC) in 1995 to supervise the implementation of its recommendations. It was replaced with the Strategic Forum for Construction in 2001, headed by Sir John Egan, author of the Egan Report *Rethinking Construction* in 1998. These reports were not isolated and followed a succession of them throughout the 20th century that considered how to deal with issues related to waste and low productivity in the UK, including *Reaching for the Skies* in 1934, the *Simon Report* in 1944 and the *Banwell Report* in 1967. In his report, Egan (1998) suggested that the industry as a whole is underachieving and asked for dramatic improvements. He proposed that this would be possible if the focus of all efforts were on delivering the value that customers need and if the industry was prepared to challenge the waste and poor quality arising from existing management structures (Designing Buildings 2023).

The report identified five key drivers of change:

1) committed leadership
2) a focus on the customer
3) integrated processes and teams
4) a quality-driven agenda
5) commitment to people

and it proposed:

1) an integrated project process
2) decent and safe working conditions
3) improved management and supervisory skills
4) replacing competitive tendering with long-term relationships
5) that the major public sector bodies became best practice clients

Both the drivers and proposals might appear naive. First, they are entirely viewed from a construction industry perspective, and secondly, they fall into the trap of all digitalisation proposals that there is an assumption of a standardised way of working with a stereotypical building type that can be applied across all countries. Thirdly, and perhaps most significantly, is the notion of language. The notion that words such as committed and commitment, customer, quality, decent, improved, best practice and clients are in any way meaningful and have shared understanding is foolhardy. To complicate the position even further, a number of other organisations were established following the Egan Report, such as the Reading Construction Forum, the Design Build Foundation, the Best Practices in Construction Program, the Movement for Innovation, the Local Government Task Force, Rethinking Construction and the Construction Clients' Group, among others. These organisations were generally combined into Constructing Excellence in 2003, although it is unclear as to what purpose. In parallel, and amongst a plethora of ill-defined intentions, the first wave of BIM implementation (formerly named 3D CAD) at the turn of the 21st century to overcome low construction productivity and other barriers hampering innovation in the industry, as related by Teicholz (2004). It should be noted that the support of architects in the design phase was not mentioned. In May 2008, even Sir John Egan declared that, after 10 years, the results obtained were quite irregular and certainly not even close to the improvement that could be achieved or expected. (Designing Buildings 2023). In 2007, BS 1192 *Collaborative production of architectural, engineering and construction information* had established a method for managing the production, distribution and quality of construction information. This included information on construction produced with computer-aided design (CAD) systems. It applied to those involved in the preparation and use of construction information during the design, construction, operation and decommissioning of projects, as well as establishing common methodologies for naming, classifying, layering and exchanging data when setting up projects that involve collaborative working as well as setting out roles and responsibilities. How much this language means to people trying to design and construct buildings is highly debatable. The launch of the UK Government Construction Strategy in May 2011 brought about a political commitment from the UK Government to drive change and transformation within the industry. It was developed by the Efficiency and Reform Group of the Cabinet Office and the Construction Sector Unit of BIS, working closely with Infrastructure UK, although most people involved in the development of the built environment knew very little about what these groups actually did. It was a coalition government riven by division and compromise, and prone to making unrealistic pledges, before the next general election. The major aim was to show the public that it could reduce the cost of construction in the public sector by up to 20% by the end of the term of the current Parliament. As with every government since, it needed to promise growth by setting ambitious and hopeful targets to be met by 2025 of cutting initial and operational costs by 33%, reducing the overall time from inception to completion by 50%, reducing carbon emissions by 50%, and improving exports by 50%. This was set against the claim that public sector design and construction represented 40% of the 110 billion pounds per year spent on construction in the United Kingdom. In fact, only a small proportion of new buildings were constructed by the public sector, and the government failed to exploit the potential for public procurement of construction and infrastructure projects. It actually established measures by which the government could make better deals with the private sector, which then became the beneficiaries. Amidst notable criticism, there were suggestions that the United Kingdom did not get full value from public sector construction. With little evidence, there seemed to be broad consensus that the construction industry was underperforming. This is particularly ironic given the substantial and high-profile failures in the IT industry. Other criticisms about the public sector were largely irrelevant, as government had ensured that the private sector was the predominant developer. To politicise the situation even further, six task forces were established:

- Procurement/Lean Client
- Standards/Lean Supply
- FM/Soft Landings
- Data and Benchmarking
- BIM Strategy
- Performance Management

It is totally unclear as to their purpose, but fragmentation of any kind of strategy would ensue. Moreover, the use of metaphors, such as soft landings, is another example of the use of language to reduce understanding. The construction industry had convinced government that BIM was the only way to manage data, but there was little consideration of how BIM implementation would be conducted. There was also criticism of the growing use of large contracts and long-term framework agreements. This kind of approach had a poor past record and had been unable to deal with the flexibility required for different building types in a variety of places. The size of the saving planned was also questioned, as the industry had already suffered from the latest recession. Never short of acronyms to abstract language even further, PAS 1192-2:2013 was introduced as a specification to manage information during the capital/delivery phase of a construction project using BIM. PAS stands for publicly available specifications. They are standards, specifications, codes of practices or quick guidelines developed by sponsoring organisations to meet an immediate market need following the directives set out by the British Standards Institution. It would be interesting to discover who the sponsoring organisations might be and who benefits from them. There is certainly an implication of standardising the design of buildings, and satisfying an immediate market need is never a good objective when considering the built environment. The specifications were commissioned by the Construction Industry Council as part of its response to the UK Government Construction Strategy, which required *fully collaborative 3D BIM as a minimum by 2016*, with all project and asset information, documentation and data being electronic. This includes the much-publicised requirement for BIM Level 2 in public projects, which is a managed 3D environment with data attached but devised in separate discipline models. PAS 1192-2:2013 focuses specifically on project delivery. Its application starts with the statement of need and works through the stages of the information delivery cycle, culminating in an as-constructed asset information model. The language is property industry-oriented and does not really relate to users or the public. The approach adopted is described as beginning with the end in mind, identifying the downstream uses of information at the outset, and ensuring its reuse during the whole building lifecycle. Throughout this process, it advocates the adoption of collaborative practices and lean construction techniques. The discussions are also about electronic data rather than building elements. As Pavan et al. (2022: 30) note, "With the BS 1192:2007, in the UK, the concept of the common data environment in construction is born, then resumed in a more digital sense in the PAS 1192-2: 2013." Should be italics without inverted commas In the context of constant change, in 2018, BS 1192 was replaced with the international standard ISO BS EN ISO 19650, as were PAS 1192-2 and PAS 1192-3.

ISO 19650 and the Collaborative Production of Information

ISO 19650 is a series of international standards. It defines the collaborative processes for effectively managing information throughout the delivery and operational phase of assets when BIM is being used. The basis is BS 1192. The outset included the collaborative production of information as defined in BS 1192 and management of information during the delivery and operational phases of the assets provided in PAS 1192-2 and PAS 1192-3, respectively. Currently, ISO 19650 comprises six parts, the first two being the most significant:

- Part 1: concepts and principles – sets out the concepts and principles recommended for developing processes and managing information throughout the lifecycle of any built asset (ISO BS EN 19650-1:2018).
- Part 2: delivery phase of the assets – defines the development processes and management of information during the development phase, in which the asset is designed, built and commissioned (ISO BS EN 19650-2:2018).

In Part 1, standards related to ISO 19650 application are proposed, thus covering all the asset life cycle as well as all those involved in the asset life cycle, such as the asset owner and operator, the client, the asset manager, the design team, the construction team, equipment manufacturer, technical specialist, regulatory authority, investor, insurer and user. These roles imply a particular type of building, on a large scale, in an IT-enabled society. Even within those limited parameters, the roles are occupied by people from a great variety of professional backgrounds, with distinctly different objectives. There is no mention of a shared language, although considerable work would need to be undertaken to engender common understanding.

Part 2 provides guidance on how to manage information during the design and construction phases. Counter-intuitively, the standard denominates these phases as asset production and delivery. It contemplates the production and delivery macroprocesses, comprising three phases: tendering, planning and production and eight stages. Everything that can be constructed is called a physical asset, whether it be a building, a building complex, or parts of it, such as an electrical system or even an electric socket. All of these physical assets can be duplicated with a corresponding virtual asset, which will be denominated in its digital twin. There are three key components that allow coordinating and structuring of the information of a virtual asset – a graphical representation in the form of one or more graphical models; a non-graphical representation in the form of asset data, with physical properties and performance and an auditable record in the form of documentation. The virtual information model developed during the design phase is called the project information model, which is constituted by the federated BIM model. The virtual information model developed during the construction phase is termed physical asset construction and is known as an asset information model. This is a complex set of digital representations aimed primarily at managing data. Linguistically, it appears at some distance from the vocabulary of form, space and structure, which is at the heart of architecture and building design.

BIM Execution Plan

This sets out to clarify functions and responsibilities, standards to be applied and procedures to be followed in the design process when a BIM is adopted. It is essential to understand that each design, with its own characteristics, in terms of size, complexity, intended use, and so on, demands its own execution plan. Therefore, even if there is a temptation to produce similar designs, everyone should be treated on their own merits. According to item 3.13.1 of ISO 19650 Part 2 (ISO BS EN ISO 19650-2:2018), an execution plan is:

- prepared between designers and contractors to define how the management of information aspects of the project will be executed
- defines functions, responsibilities, standards and procedures to be applied to the design process
- a contract document and a living management instrument
- applicable at the design level, and each design has its own plan.

Figure 13.3 shows the BIM Execution Plan in relation to the management of information process according to ISO 19650 – Part 2. The elements for establishing the BIM Execution Plan are organised and described in Figure 13.4. The language of these process charts is straightforward and understandable. Nevertheless, they are still directed towards information management and building production.

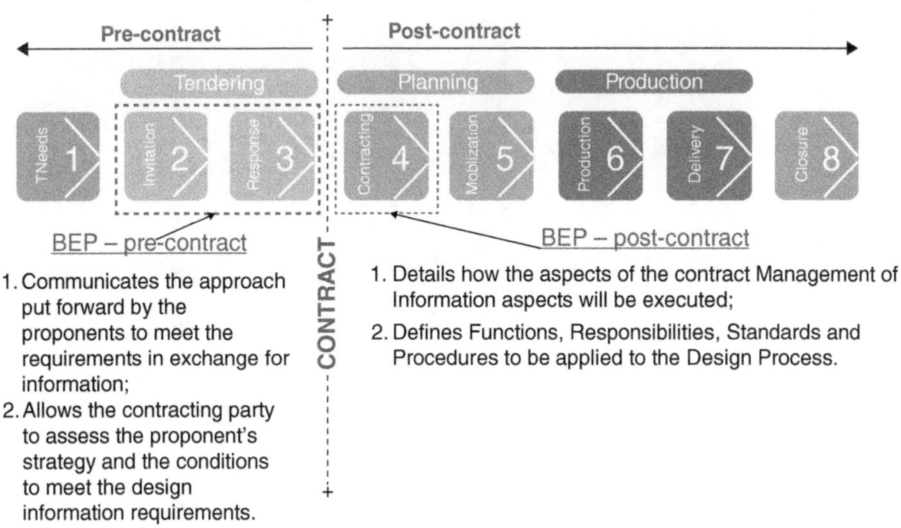

Figure 13.3 BIM execution plan: pre-contract and post-contract.

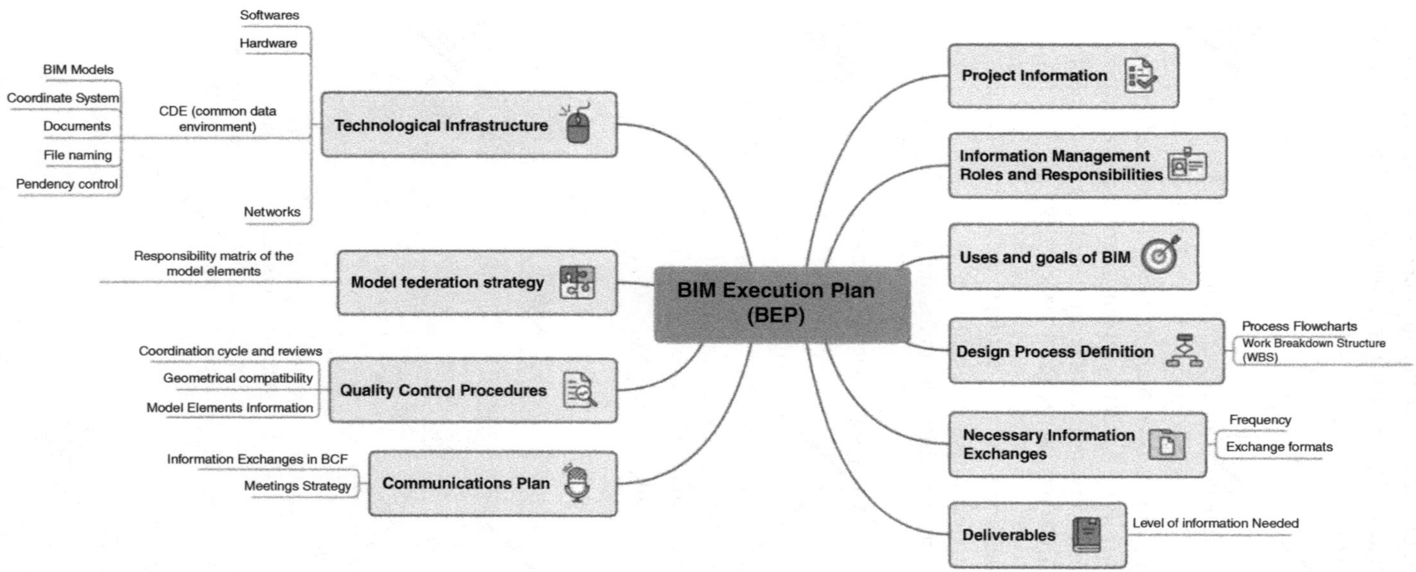

Figure 13.4 BIM execution plan: elements and structure.

Common Data Environment

The architecture, engineering, construction and operation market comprises workflows at different levels of complexity, whereby the supply chain agents must collaborate, oriented by a common set of data on the development. It is essential for these data to be available to any of the agents involved whenever necessary. Therefore, the exchange, storage and sharing mechanisms are fundamental to follow the evolution of the development throughout its lifecycle up to the delivery of the physical asset and when the client's requirements are met. To allow these collaborative workflows to materialise, the production and sharing of information containers occur in one or more common data environments. In terms of BIM, a common data environment is an "agreed source of Information for any given project or asset, for collecting, managing and disseminating each information container through a managed process" (BIM Initiative 2023). It is a server-based or cloud-based technology with database management, transmittal, issue tracking and related capabilities that support the workflow and organise the management of information throughout the lifecycle of an asset by information containers. Quality inspection is used to manage the maturity and the reliability of the information supplied, acting as the foundation for a well-structured form of collaboration among all the stakeholders. Centralising data storage allows for reducing the risk of redundancy and ensuring the availability of updated data at any time. Furthermore, a common data environment allows for a higher rate of information reuse, simplifies its aggregation and simultaneously acts as a central repository for information containers, communications and records of actions. ISO BS EN 19650-2:2018 clarifies that it can be implemented in different ways and by applying different technologies. Moreover, the design information and an asset may be in multiple sites, i.e. they can consist of a technological platform or the sum of different platforms that allow collaborating, managing, recording, and ensuring the traceability of information transferred. It is therefore important to establish the desirable functionalities and how single or multiple platforms can support the early design stage. The functionalities are indicated by a conceptual structure (see Figure 13.5).

The three main elements are as follows:

- Management of Information Containers: storage and control of the processes for exchanging documentation, data and BIM models
- Process Management: control of the processes for verifying, approving, or rejecting information containers by the changes in status and revision
- Management of Communications: control of the history, traceability of comments and requests for information, bonded or not to documents and to BIM models. This management can be performed by using open standards, such as the BIM collaboration format

As might be expected, the vocabulary for this managed process is both market-driven and set in computerised techniques. As previously discovered, the language does not relate to the established ways of presenting the design and construction process or custodianship of the built environment.

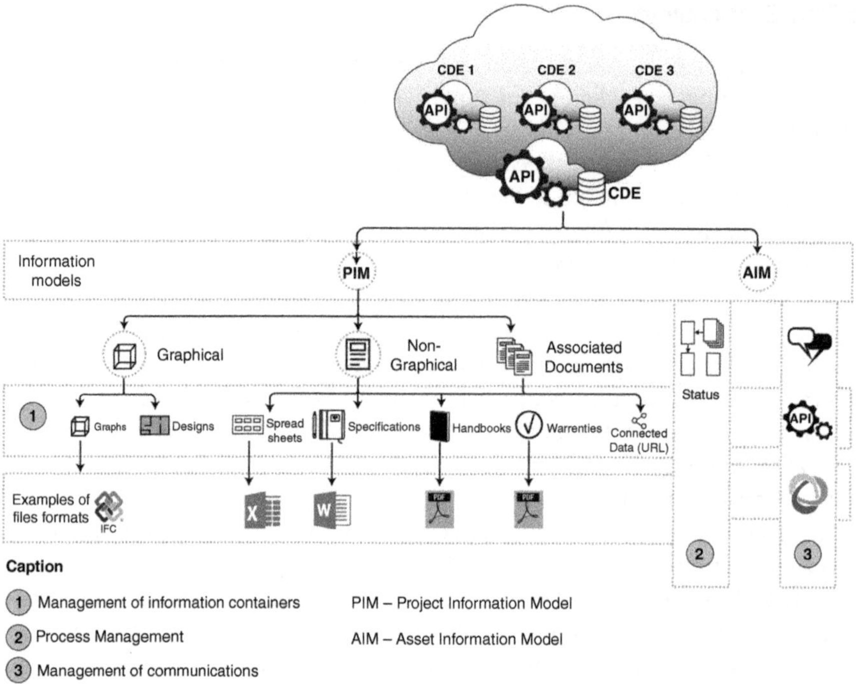

Figure 13.5 Common data elements: conceptual structure.

Experience of Countries Adaptation to Digital Construction

The reality can be very different in some countries compared to others. Nevertheless, a common issue has been found in several countries on four continents. According to the research of Cheng and Lu (2015: 465), "The public sector plays a primary role in leading industry towards BIM adoption. In some nations, the public sector is the major driver of BIM adoption. Undoubtedly, support and demonstrations by the public sector are important regarding BIM development." However, it may not be as simple as that. Additionally, the language is transformed by standards, manuals and guides, which play the role of introducing a new vocabulary of concepts, processes and rules following the digital transformation. In comparing the US and the UK, Lea et al. (2015: 513) found that "the construction industry is still trying to find its feet with standardising processes and documentation to reap the inherent benefits of BIM. Although, the adoption of industry standards and the introduction of project planning measures have been identified as critical for the successful delivery of BIM, few companies have adopted them to date". Liu et al. (2015: 168) concluded that the Chinese architecture, engineering and construction (AEC) industry is in an early stage of promotion of BIM implementation, and that "China has not published a formal national BIM standard, and the shortage of professionals in China is another main barrier to BIM implementation." The authors have presented the results of a national survey where 92.63% of respondents believe that governments should play a very important role in the BIM implementation process and give it due recognition, indicating that respondents had expectations of government support, and critical barriers were identified as the high cost of application, lack of national standards, and lack of skilled personnel. According to Jasiński (2021), in Poland, while BIM technology gives access to large-scale projects, higher rates of exploring BIM potential are not yet observed, and the promise of using BIM throughout the whole building lifecycle is still unaccomplished. Owners, developers and contractors are rarely involved in BIM model development and data implementation; instead, they are typically using the BIM model produced by the designers in a passive way. For Godager et al. (2022), in Norway, the entry of ISO 19650 raises new challenges related to the legal and contractual consequences of digitalising information management. According to the authors, where strong silos are usual, many current projects are carried out with contract forms that prevent the potential of the ISO 19650 series from being fully utilised, and the same conservatism applies to barriers to the traceability of products used in assets. It means that it is a challenge to transition from analogue to digital processes because of different maturities in different parts of the industry, stressing the fact that technological innovation requires parallel innovations in processes and policies related to the legal and contractual aspects of construction. In Australia and New Zealand, (Zima and Mitera-Kiełbasa 2022: 8) observe that "surveys show that the public is aware of BIM and is becoming more widely used over time. However, awareness does not mean application. The discovery of the benefits of BIM adaptation resulted in the organisation of national committees to help implement it. Since then, several BIM guidelines have been developed in both countries." Should be italics without inverted commas Arrotéia et al. (2021) analysed the barriers to BIM adoption in Brazil and found that BIM might be considered efficient in the design phase but flawed in the construction phase, although the evidence suggests the converse. Nevertheless, digital technology has improved the information management of the design itself and its interface with the construction phase. A more collaborative work environment has been noticed due to the effective improvement of communication between the design and construction teams on site. A universally identified issue is the capability of clients to fulfil the extent of their readiness to lead a BIM implementation process (Dakhil et al. 2019). Although a convincing definition of client is difficult to achieve, the need for anyone not directly engaged with the design and construction process to improve their ability to understand the processes and operations of BIM is a little perplexing. All these limitations to the application of BIM lead to several factors. Perhaps the ones that should be extracted here are that BIM is not applicable to all types of projects and in all countries. It may not even be applicable to all stages of the design process. In this context, it should be particularly noted that the language is highly technical, based on the computer industry rather than design and construction, difficult to penetrate and induces multiple fragmented systems.

Conclusion

It is argued that BIM is only about thirty years old, which is a very short time in the overall history of design and construction. This brief period, despite the high speed of development, leads to the conclusion that not even technology, its processes and language, as conceived by Benveniste (1971), are mature enough or consolidated. The language of digital technologies has been altered and evolved as a result of all the methods and standards that have emerged and will emerge in the next years. Certainly, the plethora of different arrangements has greatly complicated a language that was not particularly intuitive in the first instance. It is observed that ISO 19650 internationalised principles, standards, processes and subsequently the language, which expands the signifieds and creates new signifiers, introduces complex expressions and provides a vast glossary of terms that would need to be understood to engage with it. Despite the chaos that has been created, particularly induced by the intensity of BIM dissemination and its standards, the universalisation process is probably inevitable if working collaboratively at a global scale is deemed appropriate. The final and still unanswered issue for reflection is whether the ISO 19650 will close the evolution cycle of the languages and bring convergence to a common understanding. It is very challenging to spread the new vocabulary among all the stakeholders of a very fragmented industry. This is not a trivial issue, for the very task force in charge of elaborating ISO 19650 from PAS 1192 strongly recommends that each country adopt a national annex to adapt the terms to their building culture and to the existing legal framework. According to Zima and Mitera-Kiełbasa (2022), after several parts of the international standard ISO 19650 for BIM were published, some countries have introduced local annexes, e.g. in Hong Kong and Canada, and supplementary manuals were written. Nonetheless, it seems naive to assert that, from a succession of standards, various reports and a number of task forces and plans, it is possible to reach frictionless convergence or even whether this convergence is achievable. BIM technology is disruptive to the design and construction process, and its evolution shows that much is still to be resolved. New signifieds to the signifier may be a start. Alternatively, it may be concluded by many participants that the advantages of learning a new language do not match the amount of effort required.

References

Arrotéia, A.V., Freitas, R.C., and Melhado, S.B. (2021). Barriers to BIM adoption in Brazil. *Frontiers in Built Environment*, 7, 520154. https://doi.org/10.3389/fbuil.2021.520154.

Barthes, R. (1964). Éléments de sémiologie. *Communications*, 4(1964), 91–135. https://doi.org/10.3406/comm.1964.1029.

Benveniste, E. (1971). *Problems in General Linguistics* (trans. M. E. Meek). Coral Gables: University of Miami Press.

BIM Initiative. (2023). *The BIM Dictionary*. https://bimdictionary.com/en/common-data-environment/2 (accessed 3 December 2023).

Cheng, J.C. and Lu, Q. (2015). A review of the efforts and roles of the public sector for BIM adoption worldwide. *Journal of Information Technology in Construction*, 20, 442–478.

Cohesive BIM Wiki. (2023). PAS 1192-2. https://www.designingbuildings.co.uk/wiki/PAS_1192-2 (accessed 3 December 2023).

Dakhil, A., Underwood, J., Al Shawi, M. (2019). Critical success competencies for the BIM implementation process: UK construction clients, *Journal of Information Technology in Construction*, 24, 80–94. https://www.itcon.org/2019/5.

De Saussure, F. (2011). *Course in General Linguistics* (ed. P. Meisel and H. Saussy, trans. W. Baskin). New York: Columbia University Press.

Designing Buildings. (2023). Egan report rethinking construction. https://www.designingbuildings.co.uk/Egan_Report_Rethinking_Construction (accessed 17 December 2023).

Egan, J. (1998). *Rethinking Construction*. Report of the Construction Task Force to the Deputy Prime Minister, John Prescott, on the scope for improving the quality and efficiency of UK construction.

Fernie, S., Leiringer, R., and Thorpe, T. (2006). Rethinking change in construction: a critical perspective. *Building Research and Information,* 34(2), 91–103.

Fu, C., Aouad, G., Ponting, A.M., Lee, A., and Wu, S. (2004). IFC implementation in lifecycle costing. *Journal of Harbin Institute of Technology (New Series),* 11(4), 437–441.

Gaspar, J. (2019). The meaning attributed to BIM over time. Master thesis. University of Campinas. https://doi.org/10.47749/T/UNICAMP.2019.1096405

Godager, B., Mohn, K., Merschbrock, C., Klakegg, O.J. and Huang, L. (2022). Towards an improved framework for enterprise BIM: the role of ISO 19650. *Journal of Information Technology in Construction,* 27, 1075–1103. https://doi.org/10.36680/j.itcon.2022.053.

ISO BS EN 19650-1. (2018). *Organization and Digitization of Information About Buildings and Civil Engineering Works, Including Building Information Modelling (BIM) – Information Management Using Building Information Modelling – Part 1: Concepts and Principles.* ISO, Geneva, Switzerland.

ISO BS EN 19650-2 (2018). *Organization and Digitization of Information About Buildings and Civil Engineering Works, Including Building Information Modelling (BIM) – Information Management Using Building Information Modelling – Part 2: Delivery Phase of the Assets.* ISO, Geneva, Switzerland.

Jasiński, A. (2021). Impact of BIM implementation on architectural practice. *Architectural Engineering and Design Management,* 17(5–6), 447–457. https://doi.org/10.1080/17452007.2020.1854651.

Latham, M. (1994). *Constructing the Team*. Final report of the Government/industry review of procurement and contractual arrangements in the UK construction industry. London, UK: HMSO.

Lea, G., Ganah, A., Goulding, J. and Ainsworth, N. (2015). Identification and analysis of UK and US BIM standards to aid collaboration. In L. Mahdjoubi, C.A. Brebbia and R. Laing (eds.) *Building Information Modelling (BIM) in Design, Construction and Operations,* Vol. 1. WIT Press. 505–516. https://doi.org/10.2495/bim150411.

Liu, S., Xie, B., Tivendal, L., and Liu, C. (2015). Critical barriers to BIM implementation in the AEC industry. *International Journal of Marketing Studies,* 7(6), 162–171. http://www.ccsenet.org/journal/index.php/ijms/article/download/55355/29675.

Meier-Oeser, S. (2011). Medieval semiotics. In E. N. Zalta (ed.) *The Stanford Encyclopedia of Philosophy*. https://plato.stanford.edu/archives/sum2011/entries/semiotics-medieval/.

Nederveen, V.G.T. and Tolman, F. (1992). Modelling multiple views on buildings. *Automation in Construction,* 1(3), 215–224. https://doi.org/10.1016/0926-5805(92)90014-B.

Pavan, A., Caffi, V., Valra, A., Madeddu, D., Farina, D., Chiappetti, J. and Mirarchi, C. (2022). Development of BIM management system. In B. Daniotti, S. Lupica, A. Pavan and C.M. Bolognesi (eds.) *Innovative Tools and Methods Using BIM for an Efficient Renovation in Buildings*. Cham, Switzerland: Springer. https://doi.org/10.1007/978-3-031-04670-4_3.

Prins, M. and Owen, R. (2010). Integrated design and delivery solutions. *Architectural Engineering and Design Management,* 6(4), 227–231. http://dx.doi.org/10.3763/aedm.2010.

Teicholz, P. (2004). Labor productivity declines in the construction industry: causes and remedies. *AECbytes Viewpoint*, 14 April, 2004. https://aecbytes.com/viewpoint/2004/issue_4.html.

Zima, K. and Mitera-Kiełbasa, E. (2022). Level of information need for BIM models: Australia, New Zealand and ISO 19650. *Civil and Environmental Engineering Reports,* 32(4), 1–3. https://doi.org/10.2478/ceer-2022-0041.

Part VI

Final Thoughts

14
Conclusion

Bob Giddings

Northumbria University, Newcastle upon Tyne, UK

Technologies

The digital environment in architectural design and management is composed of three categories – big data, digital technologies and artificial intelligence. The first of these categories produces large quantities of design, financial and progress data in a variety of formats, including sensors, audio and video. Construction data is also available in a speedy and dynamic manner (Bilal et al. 2016). The most common digital technology is building information modelling, which has morphed into Building Information Management. It evolved from Computer-aided Design which in turn was derived from computer-aided drafting, which was seen as a digital drawing board. Other technologies have followed, including 3D printers, Augmented Reality, Virtual Reality and Mixed Reality. To date, these have largely been perceived as novel innovations, rather than mainstream process technologies. Blockchain is another global infrastructure, primarily used for recording transactions. Artificial intelligence is a means by which accumulated data from previous experiences can be held in computerised form, and multiple algorithms predict future situations as a means of more efficient decision-making. Some observers say that the programmes are acting as logical human minds, and there are concerns about them usurping human beings in decision-making.

Visualisation of proposals during periodic design reviews is clearly necessary. The three-dimensional model that is central to BIM works well in this respect. Notwithstanding, it is sophisticated, expensive and requires regular updating. Thus, it is really applicable to large and complex projects. Specific visualisation tools are also available through freeware, but they have limited functions and are generally aimed at inexperienced users. It is now suggested that machine-learning algorithms have a critical role in generating and evaluating design models (Hong et al. 2020). Digital technologies can also be used for analysis. The extent of analysis depends on the accuracy of the model geometry and supporting information. Adding building elements is resource-consuming, and therefore the budget for the model should be considered at the beginning of the project. Unfortunately, experience

Architectural Design and Management in the Digital Age: International Perspectives, First Edition.
Edited by Bob Giddings, Silvio Melhado, and Jenni Barrett.

shows that there is generally a rush to start a project, and the implications are ill-considered. They often become apparent later in the process and are disruptive at that stage. It could be argued that artificial intelligence is even more prone to partial investment, but if used correctly, it has potential for accurate automated retrieval of important project data.

The most effective use of artificial intelligence and machine-learning algorithms may be in detecting errors. Every change to the design and production information can produce clashes, duplications and omissions. The more parties there are to the process, the more likely this will occur. In addition, technologies can contribute to efficient performance attributes such as building utilisation and energy efficiency, but strong voices are required to ensure that the amenity attributes are not diminished. For example, it is claimed that algorithms can be employed to optimise architectural forms. The term optimise is interesting, as it is open to multiple interpretations.

Digital twins represent a new term for a long-established concept. Its roots are in mathematical modelling and practical equipment. The principle was that a proposed activity is modelled, the equipment constructed to replicate it and the effectiveness assessed. This is followed by a re-iterative process in which any deficiencies are fed back into the model. It is adjusted and output evaluated. If the performance is improved, the equipment is refined accordingly and the output measured. In this way, model and equipment reflect each other, and the equipment set-up is refined until improvements become minimal. Following this principle, digital twins are a bi-directional flow of information between the digital and physical environments. Data can be quantitatively and qualitatively analysed in a virtual environment, which has the characteristics of continually adapting to changes in the physical environment. Further information is then fed back into the physical environment for its enhancement. Complexity is introduced with different views regarding sub-divisions. For example, Lueth (2020) identifies 250 classifications, including levels of objects, usage and lifecycle stages. Another view by Kharche (2022) defines maturity levels as foundational, predictive, prescriptive, transformative and cognitive. A more effective way might be to adopt a holistic approach to avoid fragmentation.

Current digital twin projects are perceived as being rather limited. They are primarily focused on enhancing functional efficiency, reducing waste and improving aspects of project management. An important innovation is the notion of aligning digital twins with UN Sustainable Development Goals, as a means of broadening appeal. Nonetheless, most projects have been associated with SDGs 6 and 7, clean water and sanitation and affordable and clean energy. This is not a real advancement from the previous situation. While there are examples where any of the SDGs have been engaged in various combinations, there is a need to reflect on issues beyond functionality. In the design and management of the built environment, the principal SDG is 11 – sustainable cities and communities. Digital twins could be used to enhance inclusive and sustainable urbanisation. In terms of SDG 11, it could include recording built assets to formulate designs and budgets for upgrading or replacing them. It might identify the proportion of inadequate housing, rates of land consumption and dynamic population growth or decline in locations, to determine demand for different types of buildings and spaces. It should ensure access to adequate affordable housing, and provide access to safe, inclusive and accessible, public and green spaces. This SDG can be supported by 3 – good health and well-being; 16 – peace and justice, and the provision of strong institutions and 17 – partnerships to achieve the goals. These issues will be debated by the working commission W096 Architectural Design and Management at the CIB World Building Congress 2025.

The Human Context of Digital Work

In traditional architectural design processes, shared understanding and trust can be developed through personal contact. These teams usually work within the same organisation on series of projects, enabling organisational learning to refine the efficiency of the process and the quality of the designs. Longevity is rarely a feature of digital multi-disciplinary teams, as members tend to be brought together for the life of a single project (Nguyen and Mougenot 2022). There are two models of digital working. One is where the staff are co-located, and the other is where they are not. These are known as cyber-physical and online environments. The hybrid option of online and in person activities may require the latter to prioritise informal interaction and socially led behaviour. Where work is online only, interaction and generating trust between the participants can be more challenging, as the technological interface may act as a barrier. Thus, in digitally enabled projects, there is the constant task of ensuring the conscious management of relationships. Social ties are fundamental to team cohesiveness and, by implication, collaborative project performance (Hallstrom et al. 2021).

There are a significant number of studies that investigate why the uptake of building information management (BIM) has been slower than expected. As previously noted, it is a derivative of CAD, which itself has been shown to have benefits for creativity. Most of the uptake issues are human, cultural and social, known as collaboration blockers. The design process is composed of tacit and explicit decisions. The former represents the intuitive nature of design, while the latter involves functional information. BIM works well with explicit knowledge but constrains informal exchange associated with instinctive ways of working. Respondents to the survey by Sheppard and Barrett (2022) identified that incidental exchanges provide important feedback, and this has been lost in situations where people are engaged less in person. They also point out that the software can require a level of design data that is not always available, particularly in the early stages of the process. The debate therefore centres on whether more sophisticated technology can be developed to capture tacit knowledge and enable informal exchange, especially in an online context. Since the COVID years, online working has almost become the norm in IT-enabled societies. Nevertheless, while online meetings are suitable for progress monitoring, they lack the nuances required for design development. It has been shown that online teams do not generate cohesiveness. The continuing trend for people to increasingly work at home has led some employers to require a percentage attendance in the office. At the other end of the spectrum are virtual companies that do not have premises. This has the advantage of reducing costs on fixed assets and the flexibility of employing staff anywhere in the country or indeed the world without them needing to move. In some cases, these people have created their own offices to differentiate home from work or found their own co-workers as a response to the need for human contact.

It is now questioned whether individuals from the past, such as Frank Lloyd Wright, Le Corbusier or Mies van der Rohe, would be able to create the architecture of the future, or whether digitisation means that designs have to be generated by teams. There is also a growing notion of designs being produced from databases, with arguably a closer relationship with building production than creativity. This implies a kind of standardisation rather than recognition of the huge variety of buildings in almost an infinite number of different locations around the world. If creativity implies novelty, uniqueness and specific application, it may not suit a system very well (Bartley 2023). However, these attributes could be regarded as too unpredictable for successful buildings, and the principle of bringing together tested ideas in unfamiliar combinations may offer a more reliable route to synthesising performance and amenity attributes. A major issue is whether digital systems add to inclusion or exclusion, and this may be evident in language and vocabulary. Two opposing tendencies are evolving. One is based on monumentality, signature, no specific geography, global finance, symbols of power and starchitects. The other relates to community, context, climate change, conservation of resources and sociopolitical priorities around the globe. Exemplified by situations in Africa, there is a need to reflect on the complex history of architecture, demonstrate sensitivity to societal requirements and be innovative in creating relevant, practical and sustainable proposals to provide uniquely African solutions (Schensul and Heller 2011). There is a suggestion that digital systems entice designers to settle for what systems can produce.

It is important to examine the continued position of established practices, as younger designers move into other creative industries in addition to, or instead of, practicing architecture. There is an argument that the 19th century professions are now outdated, as evidence accumulates that young people are not only regularly changing their places of work but also moving between what might have been perceived as different careers in the past. It is also becoming debatable as to whether the architectural project should be the focus of activity, or whether it is places that should take precedence, with contributions adding to the characteristics of place. The rate of change is likely to accelerate in future but should not be represented as replacing buildings but rather through better reuse of what already exists. Although CAD has offered some inspirational forms and spaces, they tend to be related to individual buildings rather than places. Research projects with students have shown that analogue and digital actions do not need to be mutually exclusive. There are some stages of the process that benefit from one or the other, while interaction provides another level of value to design quality. Thus, instead of a doctrine that suggests analogue belongs to the past and digital to the future – flexibility and the acceptance of both having a role, will help to enhance the heterogeneity and design quality of places throughout the world.

There is an assertion that technology has the power to enable inclusivity and allow participation for people from more diverse backgrounds. Nevertheless, there are issues. If people struggle to engage in person, it may be more difficult for them to be heard digitally. Also, it may be an environment where some people can hide. Inclusion strategies could be problematic to enact if people are not together. In addition, reports about mental health as a result of isolation are disturbing. Gender and ethnically diverse organisations are shown to be more productive than others (McKinsey & Company 2020). Yet in architectural practice, despite its image as an attractive and prestigious profession, a rising number of employees are leaving due to its toxic culture and employee burnout. A notable proportion are putting significant effort into changing who they are to fit into the workplace. There should be a coordinated effort to welcome differences. Knowledge from a collection of personal lived experiences is a great resource but is mostly not used in organisational learning for architectural practice.

An advantage of digital systems such as BIM is that they obviate the need for junior architects to be involved in repetitive tasks such as scheduling. Provided this is not regarded by employers as an opportunity to down-size the staff, it releases them to consider higher-level issues. Nevertheless, they need to feel part of a team, and the learning experience may be less effective in the digital arena than in person. It should also be a two-way process, through transformational leadership, in which their contributions and needs are discussed. Digitalisation can offer support in the workplace, but personal relationships are necessary. There are organisations that provide coaching and even therapy, and this appears to be very forward-looking. At a time of development in intelligent machines, it is essential to guard against insidious bias, but the architecture profession does not seem to be involved in anti-gender bias tools. In post-colonial countries, national architectural languages will take considerable effort to develop, especially in digital form. More probable, in the drive for world-class investments, systems such as BIM are likely to produce a kind of western-style international language (Mbhokodo 2012). There is still a notion of architecture as an elitist profession, with power structures that are unlikely to be challenged by existing concepts of digitalisation. In the future, it may encourage inclusion, equity and diversity, but it will not happen automatically and will require flexibility in architectural organisation. Moreover, in some parts of the world, the cost of sophisticated equipment and its regular updating is unaffordable. Even regular access to wifi and electricity may be challenging. Therefore, simplicity, less expense and coping strategies for poor infrastructure are essential. In other parts of the world, there may be disquiet about the over-reliance on machines, especially that artificial intelligence may dominate people and the design of their environment. Alternatively, it may create a positive, inclusive situation that can confront existing power structures.

For the potential of digital architecture to reach disadvantaged and disenfranchised people, it is clear that they need to be able to access the technologies. However, where designers and communities are not within the same geographical space, this may be difficult to achieve. Alternatively, some of the locations are remote, and digital technology is ideal for those environments. In a selection of different circumstances, one technology may be examined for each case. In emergency response to natural disasters, drone mapping can be utilised for accurate data collection in large quantities over a short period of time. It can be used to reach areas that would be a considerable challenge by other means. Yet drone technology comes with ethical issues and can induce nervousness in the authorities, as to the real purpose of the drones (Wang et al. 2021). To address these and other matters, the design team needs to introduce a set of agreed protocols. In rebuilding after damage from meteorological events, increased building structures to resist further damage might be required. This may involve materials that are not available in the country and skill sets with which local people are not familiar. An alternative could be a lighter structure with its own design software. It is really important that the software should be easy to use and that the process involve community engagement (Costanza-Chock 2020). Also, in the past, there has been conflict over the lack of technology transfer and knowledge transfer, and this aspect has to be overcome.

New build design in relief areas is often in the form of warehouses, which are required for the efficient provision of supplies. To avoid past failures resulting from lack of knowledge about local circumstances, an alternative process can engage local people. Members of the community provide data on warehouse activity, context and site. The information is coordinated by a national director, who briefs the international team that uses digital technology to produce the design. This process demonstrates collaboration and partnership between local communities and international organisations. In housing marginalised groups, a 3D-scanned environment can highlight current conditions that are unnoticed by residents due to their familiarity. The virtual house enables families to appreciate how improvements to their homes can enable them to function in a more useful and enjoyable way. This would be harder to achieve by traditional means, and all members of the family would not have the same immersive experience. However, there are issues of time and money, potentially leading to digital inequalities as each home has to be individually modelled. Digital techniques can be more efficient and effective in one-off projects. Yet, there are issues. They are owned and used by international agencies, and there is little consideration of how they could become embedded in continuing activities. There are security concerns with drones and broader social and political implications about ethics and control over data with other digital methods.

Emerging Concepts

Architectural design, and for that matter, construction, can be defined as process and product. It is therefore not surprising that standardisation and customisation can be analysed in those terms. The International Organisation for Standardisation has released more than 24,600 standards. It goes back to the post-World War Two era, when a desire to restore expert knowledge became central to developing a safe, manageable and foreseeable society. This led to compartmentalisation of knowledge. As digitisation was introduced, this knowledge was translated into design workflows that integrate data-gathering, spatial simulation, documentation and coordination (Deutsch 2015) in the technocratisation and managerialisation of the design process. Architectural practices in the global arena and in IT-enabled countries, now employ professionals for their proficiency in manipulating digital tools, almost in preference to design abilities. This implies a change in the architect's role and debates about whether architectural education needs to adapt. Conversely, some architects argue that BIM causes disruptions in the design process (Holzer 2016), especially in the important iterative and re-iterative activities. Creativity is a non-linear process that appears in contrast to digital standardisation. Parametric workflows involve command sequences that control and rationalise the design process. Customisation is achieved by generating additional activities with software plug-ins. Thus, there is a case where parametricism enables architects to break out of monotonous standardisation and bring back diversity and variety. However, plug-ins to enable customisation through experimentation and iteration extend the design period, and the outcomes will involve more specialised building components and their assembly. This will require larger budgets. It may therefore be that most architects are constrained by efficiency and cost control that demands standardisation, whereas starchitects are permitted to exercise customisation.

There has been a long-held desire to achieve customisation through various combinations of standard components. Nonetheless, this approach does not achieve real variety, nor is it especially adaptable in terms of size and location. Moreover, attempts at using a limited number of components in a series of buildings, have actually resulted in multiple components being produced, as different relationships are discovered in each building. Repetition of design will create economies of scale and arguably high-quality technical products. Even so, it should be noted that in the past, there has been considerable repetition of errors through system building and factory production. An alternative to detailed design and construction that is based on standardised components would be materials and building elements within modular coordination. It was one of the objectives of metrication, which never really reached its full potential. One of the advantages of digitisation is that it can generate a framework for modular coordination. It has become clear that digitisation, and BIM in particular, is not specifically aimed at design creativity and improving the character of the built environment. While the privileged may be able to use it in that way, for most situations, it is intended to save resources, time, cost and carbon emissions.

The original objective of CAD was to enable the human designer and computer to work together (Ross 1960). Digitisation introduced new vocabulary, which the operator had to learn. As each digital tool has its own language, it is necessary to understand multiple languages. The term operator is interesting, as it is a different interpretation of the way that architects work. Perhaps the essential role of the computer is to formalise designs that have already been conceived by informal means. In that way, its logical systems can be used to reduce errors and omissions. This kind of communication can be viewed as instructive, as the machine responds to human intention. An alternative approach could be identified as questioning, where the architect defines the scope and the machine provides suggested answers. This has the feeling of a linear process that relies on how comprehensive and balanced the statements of scope might be. It may also require the digital systems to possess artificial intelligence. A more developed process might be one of ongoing instruction and questioning, with analogue and digital inputs at different stages.

The complexity of buildings and the built environment requires substantial quantities of data, and digitisation is necessary to process this data. There is a great advantage in a logical approach to analysis and synthesis that can be provided by CAD and its derivatives such as BIM. One of the issues is the intention at the beginning of the process. There is pressure for the production of a building to be done in a budget and time controlled manner. It has become apparent that the future may require buildings to be adaptable to a number of uses over relatively short periods. A hybrid approach might therefore be more effective (Carpo 2013), in which a machine-free creative design within the built environment might be the initial objective, with digitisation being employed to develop the functional attributes.

The concept of client remains complicated and contested. It is often reduced to the notion of the person or company that is paying for the development. Yet, this is a simplistic interpretation. Finance can originate from a variety of sources. It may be borrowed or on credit. The notion of a person or company spending their own money on a development could largely be a myth. Nevertheless, this is only one definition of client. Many people would argue that the users are really the client. Further, it may be the community, citizens or society. Most of these groups have a vague perception that architects use CAD, and an increasing number feel that with a few clicks, architecture can be produced with speed and little human engagement. This undermines the skills and knowledge of architects and even suggests that with digital tools, anybody can design.

Managing Digital Projects

The traditional role of the architect was to create visionary designs and enable them to be turned into reality. In different eras and locations, architects' attention to the details of a design, information for builders and involvement in the construction process varies considerably (Burr and Jones 2010). In countries that embraced the 19th century concept of professionalism, it formalised the separation of architect from builder. It has been argued that previously, architects were identified as master builders, but dilettantes were never part of the construction process. Towards the end of the 20th century, architects became involved in producing increasing amounts of information for construction. This was due to the growing professionalism of building contractors, who demanded precise instructions, together with the measurement of materials and work completed, as part of cost control. In countries where this level of professionalism was not introduced, previous practices tended to continue. The expanding scale of building contractors, with some exercising global power, changed the relationship with many architects. In a number of instances, contractors occupied the central role and employed architects for various services. These changes were stressful, especially when they were required to negotiate fees, rather than applying a statutory scale. Errors ensued and confidence diminished. The introduction of digital systems such as BIM undoubtedly alleviated this situation. They enabled omissions and errors to be detected, but almost more importantly, they created logical frameworks and reduced the person-hours spent on production information.

The introduction of digitisation in design and construction has come from different directions. A relatively small number of architects received commissions that justified investment in CAD for its capability of assisting them to produce innovative forms and spaces. Technologies such as BIM were promoted by the construction industry, for the management of production information to enable efficient construction. The industry was also able to persuade corporate clients and governments that they should be used from inception to completion and beyond (Azhar et al. 2012). Architects appreciate the advantages of three-dimensional modelling in design development, and the change from physical models to virtual models, as more flexible tools. Nevertheless, they have been a little reluctant to engage with shared models for reasons that range from loss of control over the design to issues of ownership, intellectual property, insurance and other legal matters. The phenomenal digitalisation of society should also not be underestimated. The exponential growth in smartphones, tablets and home computers has pervaded the way that everybody communicates and exchanges information. It has become central to professions and industries, so it would be inconceivable that design and construction would be an exception. Nonetheless, this may present challenges for developing countries, where limited resources can impede their ability to keep pace with developed nations. Efficient processes depend on compatibility of technology. The constant evolution of software tools can leave some individuals and organisations behind.

Roles, obligations, responsibilities and compensation can be defined by contracts, and it has already been shown that digital projects are changing organisational boundaries (Assaad et al. 2020; Winfield 2022). Contractual arrangements tend to be reactive and respond to the changes brought about by rapidly evolving digital delivery, rather than influencing digital practices. The basic function of a contract to deliver goods and services within a framework agreed upon by the parties remains unaltered, while the interface between the digital and physical worlds is increasingly blurred and the transfer of information has become a major issue. Procurement is a substantial factor in determining contract terms. A move away from traditional procurement has encouraged partial services. This kind of approach requires transfer of information, which could include a design scheme, from one party to another. The systemised nature of digital projects simplifies transfer, along with the ease and speed of actually sending material electronically. This may induce fragmentation in the process and lack of coherence in the development of a project. Alternatively, digital practice can dissolve contractual barriers.

Integrated digital project delivery is very much a property industry perspective, with a vocabulary that includes information planning and exchange and the lifecycle of a built asset. This may exclude other parts of the built environment around the world. A distinctly heterogeneous outlook to its management and contracts is therefore required to represent diversity. Integrated digital practice does impose new functions, roles and responsibilities, and there are suggestions that IT consultants might be brought into the process where traditional project stakeholders lack digital maturity. As the balance swings from the production of new buildings to the adaptability of the built environment, emphasis on IT feels increasingly abstract to the realities of place. Knowledge of buildings and spaces should perhaps take precedence over digital enablement, and technology should serve rather than lead. Procurement methods and contracts need to reflect the multiplicity of circumstances. While the process of awarding contracts to the least compliant bidder, especially those with control over the design, seems to be diminishing (Mosey et al. 2016), there are still examples of value engineering and lack of investment in the long-term enhancement of places. Ownership of digital systems and models have previously been discussed, as has intellectual property rights and ambiguity in the identification of clients. Contracts can therefore become very complicated in terms of project integration and may not be applicable beyond single-building development. Regardless of digital capabilities, people need to remain at the centre of activity. It is interesting that those who promote the exclusive use of digital systems regard social and cultural positions as barriers, whereas they could be beneficial. Therefore, the proportion of digitalisation needs to be adapted to conditions, and contracts must be made equally flexible.

Generally, digital technology fosters teams and design responsibilities that are more network-based and less hierarchically driven by professional status. Disrupting traditional ways of working, it has generated variable team structures and sometimes contested responsibilities that are yet to be fully resolved. Nevertheless, integrated project design and delivery can enhance collaboration and replace established adversarial relationships. Teams are becoming the main way of organising design and project delivery, and leadership style is crucial. Where BIM managers, for example, are included, they will need to find an appropriate place in the decision-making structure, rather than duplicating or usurping the work of architects. It is clear that teams are more effective if they represent diversification. This is not easy to achieve, as people feel more comfortable when surrounded by their own kind and like to hear their own beliefs being fed back to them. Sustainability has been much neglected in the development of many building projects. It is now argued that, especially in complex projects, design teams require specific members who would promote economic, social and environmental sustainability.

There are some positive moves in the specialised uses of technology, such as BIM. For example, ancient but still developing cities in Turkey are benefitting from it for the management of their built heritage (Bastem and Cekmis 2022). It is also a basis for identifying reusable elements in the deconstruction of redundant buildings (van den Berg et al. 2021). Selective design for manufacture and assembly can be a helpful strategy for integration of design and build (Gao et al. 2020). The word selective is pivotal, as previous versions of factory production have resulted in negative appearance of the finished buildings as well as technical failures. This incarnation should focus on non-visual but functional aspects of the design. On the other end of the spectrum, well-known award-winning architects have stated publicly that they do not wish to engage with digital technology at all. Others have written that a non-digital option may not be available for architects designing future complex projects. Therein lies one of the major issues. By their nature, these complex projects will have extensive design teams, in which coordination and information flow are critical to effective management. It should therefore be expressed once again that the evolution of built environments around the world involves a huge variety of buildings that can be designed with various tools. Some tools may be analogue, some will be digital, and architects may use different tools to suit the location and nature of intervention in the built environment, or even within one building design. Whatever the tools, they are not a replacement for knowledge and skills. Moreover, while there are very valuable contributions that can be made by design team members, architectural education is the necessary prerequisite for designing buildings.

Language and vocabulary are fundamental to understanding and communication. Architectural language has its own vocabulary to describe elements of buildings and assist in the process and product of design. The development of computer-based technologies has introduced a whole new vocabulary into the process. Students often struggle to assimilate both the language of architectural design and digital technologies. The latter is creating a diverging vocabulary, which may be hindering a shared understanding. As digital technologies evolve rapidly, so do their languages. This can create clique cultures that exclude those who are not keeping pace. This phenomenon may also invade the vocabularies of everyday life. An increasing number of words and grammatical expressions move in and out of vogue, and digital technologies and their interconnectivity are often the drivers. Words can also be replaced by symbols and acronyms. These can also create inclusion and exclusion in communication, where knowledge of the symbols and acronyms assists communication, but being unable to decipher them produces a situation in which it is impossible to progress. The acronym BIM is simple, powerful and easily reported, as indicated by its presence in the chapters of this book. Nonetheless, its language seems to be deviating from the vocabulary of design and construction in the built environment; for example, the BIM Initiative (2023) identifies terms in its dictionary such as disseminating information containers through managed processes and common data elements.

Standards, manuals and guides will continue the transformation of language, with large glossaries of terms that will need to be understood by all parties to the process. Even the International Standard for Building Information Modelling (ISO 19650) presents notable challenges in spreading a new vocabulary within such a fragmented industry. It is considered to be one of the major reasons why, even in the United States and United Kingdom, few companies have adopted standards to date. In China, a formal national standard has not yet been published, and the whole BIM implementation process will require government support due to the high cost of application and lack of skilled personnel. Progress is slow in Poland (Jasiński 2021). In Norway, the adoption of ISO 19650 is experiencing many inhibitions (Godager et al. 2022). In Australia and New Zealand, national committees have been established to help implement BIM (Zima and Mitera-Kiełbasa 2022), and in Brazil, the barriers to BIM adoption have been analysed by Arrotéia et al. (2021). An unanswered issue remains as to whether ISO 9650 can address the language dilemma and bring convergence to a common understanding. This is not a trivial matter, and it seems unlikely that a simple standard can bring about this convergence. Thus, the initiative is being handed back to each country, with their varying abilities to be able to respond.

References

Arrotéia, A.V., Freitas, R.C. and Melhado, S.B. (2021). Barriers to BIM adoption in Brazil. *Frontiers in Built Environment*, 7, 520154. https://doi.org/10.3389/fbuil.2021.520154.

Assaad, R., El-adaway, I.H., El Hakea, A.H., Parker, M.J., Henderson, T.I., Salvo, C.R. and Ahmed, M.O. (2020). Contractual perspective for BIM utilization in US construction projects. *Journal of Construction Engineering and Management*, 146(12), 04020128. https://doi.org/10.1061/(ASCE)CO.1943-7862.0001927.

Azhar, S., Khalfan, M. and Maqsood, T. (2012). Building information modeling (BIM): now and beyond. *The Australasian Journal of Construction Economics and Building*, 12(4), 15–28.

Bartley, G. (2023). Creative being. *Philosophy Now*. 153. Editorial. 3. https://philosophynow.org/issues/153/Creative_Being.

Bastem, S.S. and Cekmis, A. (2022). Development of historic building information modelling: a systematic literature review. *Building Research and Information*, 50(5), 527–558.

van den Berg, M., Voordijk, H. and Adriaanse, A. (2021). BIM uses for deconstruction: an activity-theoretical perspective on reorganising end-of-life practices. *Construction Management and Economics*, 39(4), 323–339.

Bilal, M., Oyedele, L.O., Akinade, O.O., Ajayi, S.O., Alaka, H.A., Owolabi, H.A., Qadir, J., Pasha, M. and Bello, S.A. (2016). Big data architecture for construction waste analytics: a conceptual framework. *Journal of Building Engineering*, 6, 144–156. https://doi.org/10.1016/j.jobe.2016.03.002.

BIM Initiative. (2023). *The BIM Dictionary*. https://bimdictionary.com/en/common-data-environment/2 (accessed 3 December 2023).

Burr, K.L. and Jones, C.B. (2010). The role of the architect: changes of the past, practices of the present, and indications of the future. *International Journal of Construction Education and Research*, 6(2), 122–138. https://doi.org/10.1080/15578771.2010.482878.

Carpo, M. (2013). Twenty years of digital design. In M. Carpo (ed.) *The Digital Turn in Architecture 1992–2012*. Hoboken, NJ: John Wiley and Sons. 8–14.

Costanza-Chock, S. (2020). *Design Justice: Community-Led Practices to Build the Worlds We Need*. Cambridge, MA: The MIT Press.

Deutsch, R. (2015). *Data-Driven Design and Construction*. Hoboken, NJ: John Wiley and Sons.

Gao, S., Jin, R. and Lu, W. (2020). Design for manufacture and assembly in construction: a review. *Building Research and Information*, 48(5), 538–550.

Godager, B., Mohn, K., Merschbrock, C., Klakegg, O.J. and Huang, L. (2022). Towards an improved framework for enterprise BIM: the role of ISO 19650. *Journal of Information Technology in Construction*, 27, 1075–1103. https://doi.org/10.36680/j.itcon.2022.053.

Hallstrom, A., Bosch-Sijtsema, P., Polete, L., Rempling, R. and Karlsson, M. (2021). The role of social ties in collaborative project networks: a tale of two construction cases. *Construction Management and Economics*, 39(9), 723–738.

Holzer, D. (2016). *The BIM Manager's Handbook: Guidance for Professions in Architecture, Engineering and Construction*. Chichester: John Wiley and Sons.

Hong, T., Wang, Z., Luo, X. and Zhang, W. (2020). State-of-the-art on research and applications of machine learning in the building life cycle. *Energy and Buildings*, 212, 109831. https://doi.org/10.1016/j.enbuild.2020.109831.

Jasiński, A. (2021). Impact of BIM implementation on architectural practice. *Architectural Engineering and Design Management*, 17(5–6), 447–457. https://doi.org/10.1080/17452007.2020.1854651.

Kharche, V. (2022). *Digital Twin Maturity*. https://www.linkedin.com/pulse/digital-twin-maturity-vinayak-kharche/ (accessed 10 December 2023).

Lueth, K.L., (2020). How the world's 250 digital twins compare? Same, same but different. *IoT Analytics*. https://iot-analytics.com/how-the-worlds-250-digital-twins-compare/ (accessed 10 December 2023).

Mbhokodo, I.M. (2012). Heritage, an asset for the development of the capital, in City of Tshwane. In N. Clarke and M. Kuipers (eds.) *Re-Centring Tshwane – Urban Heritage Strategies for a Resilient Capital*. Pretoria: Visual Books. 27–37. http://www.braaksma-roos.nl/wp-content/uploads/2017/11/Re-Centring-Tshwane-Urban-heritage-strategies-for-a-resilient-capital-Nicholas-Clarke-Marieke-Kuipers-Job-Roos.pdf.

McKinsey & Company. (2020). *Diversity Wins: How Inclusion Matters*. New York: McKinsey & Company.

Mosey, D., Bahram, D., Dartnell, R., Hallam, C., Howard, C., Maqbool, A., Murray, K., Rawlinson, S. and Winfield, M. (2016). Enabling BIM through procurement and contracts. *Society of Construction Law Papers D*. 192.

Nguyen, M. and Mougenot, C. (2022). A systematic review of empirical studies on multidisciplinary design collaboration: findings, methods, and challenges. *Design Studies*, 81, 101–120.

Ross, D.T. (1960). *Computer-Aided Design: A Statement of Objectives*. Technical Memorandum 8436, MIT Electronic Systems Laboratory. Cambridge, MA: Massachusetts Institute of Technology.

Schensul, D. and Heller, P. (2011). Legacies, change and transformation in the post-apartheid city: towards an urban sociological cartography. *International Journal of Urban and Regional Research*, 35(1), 78–109. https://doi.org/10.1111/j.1468-2427.2010.00980.x.

Sheppard, J. and Barrett, J. (2022). *Building on Binary: Integrating Creative Collaboration Within Digital Design Processes in the Construction Industry*. Research Study (*unpublished*). University of Cambridge.

Wang, N., Christen, M. and Hunt, M. (2021). Ethical considerations associated with "humanitarian drones": a scoping literature review. *Science and Engineering Ethics,* 27(4), 51. https://doi.org/10.1007/s11948-021-00327-4.

Winfield, M. (2022). Legal implications of digitization in the construction industry. In M. Bolpagni, R. Gavina and D. Ribeiro (eds.) *Industry 4.0 for the Built Environment,* Vol. 20. Cham, Switzerland: Springer. 391–411. https://doi.org/10.1007/978-3-030-82430-3_17.

Zima, K. and Mitera-Kiełbasa, E. (2022). Level of information need for BIM models: Australia, New Zealand and ISO 19650. *Civil and Environmental Engineering Reports,* 32(4), 1–3. https://doi.org/10.2478/ceer-2022-0041.

Index

Architectural Design and Management in the Digital Age: International Perspectives, First Edition.
Edited by Bob Giddings, Silvio Melhado, and Jenni Barrett.